Edwin James Houston, Arthur Edwin Kennelly

Electric Heating

Edwin James Houston, Arthur Edwin Kennelly

Electric Heating

ISBN/EAN: 9783337802172

Printed in Europe, USA, Canada, Australia, Japan

Cover: Foto ©berggeist007 / pixelio.de

More available books at **www.hansebooks.com**

ELEMENTARY ELECTRO-TECHNICAL SERIES

ELECTRIC HEATING

BY

EDWIN J. HOUSTON, Ph. D., (Princeton)

AND

A. E. KENNELLY, Sc. D.

NEW YORK
THE W. J. JOHNSTON COMPANY.
253 Broadway
1895

PREFACE

IN preparing this volume on ELECTRIC HEATING, as one of a series entitled *The Elementary Electro-Technical Series*, the authors believe they are meeting a demand, that exists on the part of the general public, for reliable information respecting such matters in electricity as can be readily understood by those not especially trained in electro-technics.

The subject of electric heating is to-day attracting no little attention. The wonderful growth in electric street railways, coupled with the readiness with which the current can be applied to the heating of the cars, together with the marked efficiency of the electric air heater as an apparatus for transforming electric energy into heat energy, have, during the last decade, caused a development in electric car heating. But the growth of electric heating has by no means been limited to this particular field. The development of electric cooking apparatus has naturally attended the extensive distribution of electricity for lighting and power, and electric cooking is now taking its place with electric lighting as an adjunct to the modern house.

In the direction of the employment of powerful electric currents for heating effects, processes for electric welding, and the electrical shaping and forging of metals, are coming into commercial use, and applications are daily being made of the power of electricity in electric furnaces, either where the heating effect alone is employed, or where both heating and electrolytic effects are utilized.

CONTENTS

		PAGE
I.	INTRODUCTORY	7
II.	ELEMENTARY PRINCIPLES	30
III.	ELECTRICAL HEATING OF BARE CONDUCTORS	37
IV.	ELECTRICAL HEATING OF COVERED CONDUCTORS	69
V.	FUSE WIRES	87
VI.	ELECTRIC HEATERS	117
VII.	ELECTRIC COOKING	151
VIII.	ELECTRIC WELDING	181
IX.	ELECTRIC FURNACES	233
X.	MISCELLANEOUS APPLICATIONS OF ELECTRIC HEATING	255
	INDEX	271

ELECTRIC HEATING.

CHAPTER I.

INTRODUCTORY.

———o———

ZOROASTER, the founder of fire worship, because of the many advantages mankind derived from fire, bade his followers worship the sun as its prime and sustaining cause. Although the idolatrous doctrine of the old Persian is now entirely discredited by civilized races, yet the truth of the belief that found in the sun the source of all the thermal phenomena of the earth, still remains unchallenged. It can be shown, from a scientific point of view, that in reality, there is not one of the many ways in which man can produce heat on the earth, that cannot trace its prime cause to the sun.

Take, for example, one of the commonest methods of obtaining heat; namely, by the burning of a mass of coal. Here it is, at first sight, by no means evident, that the heat of the glowing mass was derived from the sun. In accordance with modern scientific belief, heat is no longer regarded as a kind of matter, but as a condition of matter. A hot body differs from a cold body in that the very small particles or molecules, of which it is composed, are in a state of rapid to-and-fro motions or oscillations. When a hot body grows hotter, the only effect produced, unless the body is melted or evaporated, is to increase the violence of these molecular oscillations. Could we deprive a body of all its heat its oscillations would entirely cease. In order to produce molecular or heat oscillations, energy must be expended on the body; that is, work must be done on its molecules. In other words, a hot body is a mass of matter plus a certain quantity of molecular energy. When a hot body cools, it throws off or dissipates a certain quantity of its molecular energy, and, when the heat thus thrown off is absorbed or taken in by another body, the latter thereby acquires an additional store of energy. When a pound of coal is burnt in air, the heat produced results from the mutual attractions existing between the molecules of the carbon and the molecules of the oxygen in the air; or, from what is ordinarily called their *chemical affinity*.

Unburnt coal and air possess, jointly, a store of chemical energy, having the power or potency of doing work, but actually doing no work; while coal and air after burning, no longer possess this store of chemical energy, but have acquired in its place a stock of oscillation or heat energy; *i. e.*, energy of oscillation.

Could the burning be effected in a heat-tight space, this oscillation or heat energy would be entirely confined to the interior of the chamber, but as no bodies are perfect non-conductors of heat, such a heat-tight space cannot be obtained, and some, at least, of the oscillation energy will be communicated to surrounding bodies.

A steam engine is a machine for producing mechanical energy at the expense of molecular oscillation energy. If we suppose that a pound of coal could be burned in connection with a theoretically perfect steam engine, with the necessary quantity of air, all the molecular oscillation energy developed by the combustion could be utilized by the engine, which would do an amount of work exactly equal to the amount of original chemical energy residing in the coal and air. It is known, as the result of calculation, that such an engine would be capable of doing an amount of work represented by the lifting of one pound through a height of about 2000 miles. When, therefore, a pound of coal is burnt with air, an amount of oscillation energy is developed, equal to that which would be obtained by the falling of that pound of coal from a height of about 2000 miles. Owing to a variety of circumstances, however, the best steam engines are only capable of yielding about 15 per cent. of this work.

The store of energy existing in a pound of coal was obtained from the sun's radiation during the geological past. That is to say, during the Carboniferous Age, the carbon of the coal originally existed in the earth's atmosphere combined with oxygen as gaseous carbon dioxide. For the formation of every pound of coal existing in the earth's crust a definite quantity of carbonic acid gas was dissociated, or separated into carbon and oxygen, by means of the energy of the sun's rays absorbed by the vegetation of the Carboniferous Era. In other words, the leaves of the carboniferous flora absorbed gaseous carbon dioxide from the atmosphere, and, in the delicate laboratories of the leaf, by means of the energy absorbed directly from the sun's rays, a dissociation occurred between the carbon and the oxygen. The ability, therefore, of the carbon to again recombine with oxygen in the form of gaseous carbon dioxide has been a result of energy expended on the plant and lodged in the carbon of its woody fibre. A lump of coal, therefore, is in reality a store-house of the solar heat of an early geological era.

Viewed in this light, a lump of coal can be regarded as not unlike a weight raised, say from the ground through a certain height. Suppose, for example, a pound weight be attached to a string passing over a pulley and raised to a height of 20 feet from the ground, and that, while in this position, the string of the pulley be fixed. Evidently, work has been expended in raising the pound weight, and, as a result of this work, the weight is placed in a position in which it can, at any time the string is released, fall back again to the ground, and in so doing restore the amount of work originally expended in lifting it. In the same way, a pound of coal has, by the work of the sun, been placed in a condition in which it can combine with the oxygen of the air, and burn. In so doing it must give out an amount of heat equal to that representing the sun's work upon it, amounting, as we have seen, measured in units of the earth's gravitational work, to an elevation of about 2000 miles.

But it was not only during the geological past that the solar energy was thus husbanded in the earth's crust. The sun's energy is to-day being similarly stored in all vegetable foods, and it is on this store that animals draw for their muscular and nervous energy. That is to say, all vegetable products represent chemical stores of solar energy. An animal is capable of releasing this energy in its muscles by the actual combustion of these chemical substances, after their proper assimilation in its body. Muscular activity, therefore, is but another instance of energy primarily obtained from the sun's radiation. The earth's animals are, therefore, in this sense truly children of the sun, since they thus indirectly derive their activity from that luminary.

Not only can the heat and consequent mechanical motion, which it is possible to obtain by the burning of a mass of coal, or by the assimilation and consequent oxidation of a certain quantity of food by an animal, be traced indirectly to the sun, but the same can also be shown to be true for all the other sources of mechanical energy with which we are acquainted on the earth. Take, for example, the energy delivered to a windmill from moving air, or to a water-wheel from flowing water. In the case of a windmill, the sun's heat, acting upon the air, sets up convection currents, or winds, whereby the work expended by the sun in heating the air is liberated in mass motion. In the case of a water-wheel, where a stream of water flowing from a higher to a lower level is caused to impart its energy to the wheel, the water, in reality, occupies a position corresponding to a raised weight, and is able to do work because, like the water, it is at the higher level. To what source of energy does it owe this ability to do work? Manifestly to the heat of the sun, whereby the water was raised as vapor and subsequently fell as rain on the slopes of the higher level from which it is now flowing to a lower level.

The molecules of a hot body are moving

to-and-fro at varying velocities. Some are moving faster than others; for, during their to-and-fro motions, they frequently collide, some molecules being thereby accelerated and others retarded. The average molecule, of a given mass possessing a given amount of heat, may, however, be assumed to possess on the whole, a certain average velocity of motion. It is clear, therefore, that if we could transform the molecular oscillations of a heated body into a motion of the whole mass, the body would move with a uniform velocity which would be its average molecular velocity, in the sense just described. This conception is valuable as affording a measure of the amount of heat possessed by a body. Similarly, when work is done upon a body, whereby it acquires, or is capable of acquiring, a certain velocity of motion, this motion can be represented by an agitation of the molecules in the quiescent mass of the body, the average molecular velocity corresponding to the velocity of the mass. Clearly, therefore, heat represents mechanical work, and mechanical work represents heat. Or, in other words, a certain quantity of mechanical work is capable of being expressed as a definite quantity of heat, or a certain amount of heat is capable of being expressed as a definite quantity of mechanical work, even though, in all cases, we may not, at present, possess the means whereby the actual conversion of one into the other can be effected.

For this reason a given quantity of heat can only be made to produce a certain quantity of mechanical work corresponding thereto, even though the means of conversion were so perfect that no loss should take place during the process. And similarly, a given quantity of work can only be capable of developing a fixed quantity of heat, no matter how perfect the mechanism of conversion may be.

Heat developed by electricity forms no exception to the preceding principles. As we shall see, a given quantity of electrical energy is capable of producing a fixed quantity of heat, no matter how such heat is developed. The limit to improvement in electrical heating apparatus, as in any other machinery, being such as will insure the least loss of energy during the process of conversion. As a matter of fact, electrical energy can always be completely converted into heat, although, unfortunately, the converse is not, at present, true, and heat energy cannot, therefore, be completely converted into electrical energy, but only a comparatively small fraction can be so converted.

It is a fundamental doctrine of modern science that energy is never annihilated. It apparently disappears in one form, only to reappear in another form. Thus, heat energy, or molecular motion, when disappearing as such, reappears in some other form; say, for example, as mass motion, or mechanical energy. Mechanical energy may in its turn disappear as such, to produce chemical, thermal, electromagnetic, or some other form of energy. In all cases, definite quantitative relations exist between the amounts of energy exchanged, but in every process of conversion a tendency exists whereby some of the energy assumes the form of molecular motion or heat, in which it is often impossible to again utilize or further transform it.

CHAPTER II.

ELEMENTARY PRINCIPLES.

DURING the building of a brick wall, a certain amount of work is done in raising the bricks from their position on the ground to their position in the wall. The amount of this work is definite, and is measured by the amount of force required to raise the bricks directly against the gravitational pull of the earth, multiplied by the vertical distance through which they are raised.

Care must be taken not to confuse the ideas of force and work. Force may be defined, in general, to be that which causes a body to move, or to tend to move. Work is never done by a force unless it actually produces a motion in the body on which it is acting. For example, when a brick rests on a wall, or on the ground, it is exerting a force vertically downward in virtue of the earth's gravitational pull; that is to say, it is pressing downward against the earth, with a force equal to its weight, approximately six pounds' weight, but this force is not doing work since it is not producing a motion of the brick. Work had to be done on the brick when it was raised from the ground to its position on the wall; that is, a muscular force, equal to that of six pounds' weight, had to be exerted, in order to overcome the earth's gravitational attraction on the brick and this force had to be continuously exerted while the brick was being raised through the vertical distance existing between the ground and its position in the wall. Moreover, if the brick be permitted to fall from the wall to the ground, work will be done by the brick in falling, which could be usefully employed, as, for example, in winding a clock, and the amount of this work could be represented, as before, by the weight of the brick multiplied by the distance through which it falls. This amount of work must be equal to that which was expended in lifting the brick.

In order to measure accurately the amount of mechanical work done *on* a body in raising it through a given vertical distance, or the amount of work done *by* a body in falling, reference is had to certain *units of work*. A convenient unit of work, much employed in engineering, in the United States and in England, is called the *foot-pound*, and is equal to the work done when a force equal to a pound's weight acts through a distance of one foot.

Suppose a uniform brick wall containing 1000 bricks, each weighing, say six pounds, has its top six feet from the ground. The total weight of the wall would be 6000 pounds, and the average distance through which the bricks would have to be raised, in building the wall, would be three feet, so that the amount of work necessarily expended in the building of the wall, would be that required to raise its weight through its average height, or $6000 \times 3 = 18{,}000$ foot-pounds.

The foot-pound is not employed as a unit of work in countries outside of the United States and Great Britain, nor generally in scientific writings anywhere. A unit frequently employed is called the *joule*, and is commonly used as the unit of work performed by an electric current; for, as we shall see, electric currents are capable of doing work. The value of the joule may, however, be conveniently expressed as being approximately equal to 0.738 foot-pound; or to the work done in raising a pound through nearly nine inches. Thus, the amount of work expended in the building of the brick wall just referred to, was 18,000 foot-pounds, or approximately 24,000 joules.

The brick wall referred to in the preceding paragraph might be erected by the workmen in a day, or in six days, but, when built, the amount of work done would be the same; namely, 24,400 joules. Regarding its erection from the standpoint of each workman, the rate at which each man would have to expend his energy in doing the work would be very different in the two cases, since, if he does in one day that which he would otherwise do in six days, he would clearly expend his energy at an average rate six times greater in the former case. The rate at which work is done is called *activity*, so that the average activity of the workman would be six times greater, if the wall is built in one day, than if it be built in six days.

A *unit of activity* is the *foot-pound-per-second*. As generally employed in England and America, the practical unit of activity is the average activity of a certain horse assumed as a standard. This unit of activity is called the *horse-power*, and is an activity of 550 foot-pounds-per-second. The unit of electrical activity generally used all over the world and which may, therefore, be called the *international unit of activity* is the *joule-per-second*, or the *watt*, and is equal to 0.738

ELEMENTARY PRINCIPLES.

foot-pound-per-second, or 1-746th of a horse-power, so that 746 watts are equal to one horse-power.

The engines of an Atlantic liner may develop steadily about 30,000 H. P. in driving its propellor. This represents an activity of 30,000 × 550 = 16,500,000 foot-pounds, or 8250 short-tons, lifted one foot-per-second, or one short ton lifted 8250 feet-per-second; or, expressed in watts, or joules-per-second, 22,374,000.

A laborer digging a trench will usually average an activity of only 50 watts, or 36.9 foot-pounds-per-second, during his work, so that the average activity of a laboring man may be taken as about 1-15th of a horse-power. A man frequently works, however, at an activity much greater than this, say at an activity of 100 watts, or about 1-8th horse-power, while for short periods, say for half a minute, he can sustain an activity of, perhaps, 500 watts, or even 746 watts, or one horse-power.

As we have already seen, a definite and fixed relation is maintained between the amount of heat or oscillation energy present in a unit quantity of matter, say a pound of water, and the amount of energy which must be expended on this matter in order to heat it to a given temperature. The amount of heat energy in an indefinite quantity of a body, such as water, cannot be determined from its temperature alone; we require, beside this, to know its mass. If we know its weight in pounds, and its temperature in degrees; *i.e.*, the pound-degrees, we can determine the quantity of heat energy existing in the mass. In other words, the *pound-degree* may be taken as a *heat unit*, and, since this represents a definite amount of work, this heat unit may have its value expressed either in joules or in foot-pounds. The *British heat unit*, sometimes called the *British thermal unit*, or the *B. T. U.*, is the amount of heat required to raise a pound of water one degree Fahrenheit, from 59° to 60° F. and is taken as 778 foot-pounds, or 1055 joules. The heat unit most frequently employed in countries other than the United States and Great Britain, is the amount of heat required to raise one gramme of water 1 C. This heat is called the *water-gramme-degree-centigrade*, the *lesser calorie*, or the *therm*. Expressed in foot-pounds, one lesser calorie is equal to 4.18 joules, or 3.087 foot-pounds.

The amount of work expended in heating a cubic foot of water, of approximately 62½ pounds weight, from 50° F. to the boiling point of 212° F., or through a temperature of 162° F., is approximately $62\frac{1}{2} \times 162 = 1013$ B. T. U. $= 1,069,000$ joules.

A reservoir filled with water possesses a certain store of energy, or capacity for doing work, dependent both on the amount of water it contains and on the distance through which the water is permitted to flow in escaping from the reservoir. In accordance with what has already been stated, the amount of this work can be represented by the weight of the water in pounds, multiplied by the distance in feet through which the water falls. Thus, consider a reservoir holding, say 100,000 cubic feet of water, at a mean elevation of 10 feet above a pump which fills it. The weight of the water would be approximately 6,250,000 pounds, and the amount of work required to be expended by the pump in lifting it 10 feet would be approximately 62,500,000 foot-pounds, or 84,750,000 joules. If, now, this water be permitted to escape to the pump level, in so doing it will expend just this amount of work. If the distance through which the water fell were twice as great; *i.e.*, if the pump level were 10 feet lower down, then half the quantity falling through this double distance would do the same amount of work, and, of course, to fill the reservoir through such a distance would necessitate the expenditure of twice as much work as in the former case.

Although electricity is not to be considered as a liquid, yet many of the laws which relate to its flow are similar to the laws controlling liquid flow. For example, in order to obtain a flow of water, a difference of pressure must exist, generally in the form of a difference of water level, and the direction of the current of water is from the higher to the lower pressure, or from the higher to the lower level. So, too, in order to obtain a flow of electricity, a *difference of electrical pressure or level* must exist, or, as it is commonly called, an *electromotive force*, and the direction of the electric current is assumed to be from the higher to the lower pressure, or from the higher to the lower electric level. Just as in the case of the water flow, the quantity of water is represented by some unit quantity, such as a pound, so in the case of the electric current, the quantity of electricity is represented by a *unit of electric quantity* called a *coulomb;* and, as in the case of the water, the difference of level is represented by some such unit as a foot, so in the case of the electric flow, the *difference of electric pressure or level,* is represented by a unit called the *volt*. Moreover, as the amount of work done by a given quantity of water in flowing, is equal to the quantity of water represented, say, in pounds, multiplied by the distance through which it moves in feet, the work being expressed in foot-pounds, so the amount of work done in an electric circuit, by the electric current in flowing, is equal to the quantity of electricity in coulombs, multiplied by the pressure, or the difference of electric level through which it flows, in volts. The work being expressed in *coulomb-volts*, or *joules*, a joule being equal to one coulomb-volt. In point of fact the name joule, for a unit of work, was first employed as the name of the coulomb-volt, the unit of electric work.

When a flow of 100 coulombs of electricity passes through a circuit under a pressure of 50 volts, the amount of work expended by the electric current will be 100 × 50 = 5000 joules = 3690 foot-pounds; one coulomb of electricity passing under a pressure, or through a difference of electric level, of 100 volts, will expend the same amount of work; *i. e.*, 100 joules, as 100 coulombs passing under a pressure of one volt. An electric source, such as a dynamo, or a voltaic battery, is a device for producing an electromotive force; that is, a difference of electric level or electric pressure in a circuit, just as a pump is a device for producing a difference of water level as in forcing water into a reservoir.

The activity of a reservoir, when discharging water, depends upon the quantity of water escaping per second; and, as in the case of all activity, may be expressed in foot-pounds-per-second, or in

ELEMENTARY PRINCIPLES. 35

watts. So in an electric circuit, the activity depends upon the flow of electricity per second through a given difference of electric level, or electromotive force, (abbreviated E. M. F.) and is also expressed in joules-per-second, or in watts. Thus, when 100 coulombs pass through an electric circuit under a pressure of 50 volts. a total work of 5000 joules will be done, and if this work be expended in one second, the activity during that time will be 5000 watts. If the same total flow take place steadily in 50 seconds, the flow-per-second would be 2 coulombs, and the activity, $50 \times 2 = 100$ volt-coulombs-per-second, or 100 watts.

An electric flow may be expressed in coulombs-per-second; *i. e.*, in amperes. Since an ampere is a rate of flow of one coulomb-per-second, electric activity can be expressed in *volt-coulombs-per-second*, or in *volt-amperes; i. e.*, in *watts*. A circuit in which 10 amperes is flowing under a pressure of 100 volts, is having electric energy expended in it at the rate of 100 × 10 volt-amperes, or 1000 watts, or 1 kilowatt. A *kilowatt* is the unit commonly employed in the rating of electrical machinery, since the watt is too small a unit for convenience. One kilowatt, abbreviated KW., is equal to 1.34 H.P., or, approximately, $1\frac{1}{3}$ H.P.

CHAPTER III.

ELECTRICAL HEATING OF BARE CONDUCTORS.

THE quantity of water which escapes from a reservoir in a given time depends not only on the pressure at the outlet, but also on the diameter and length of the outlet pipe. So, too, when an electric current flows through a conducting circuit, the quantity of electricity which passes per second; *i. e.*, the coulombs-per-second, or the amperes, depends not only on the pressure, or the E. M. F., but also on the length and dimensions of the conductor, as well as on the material of which the conductor is composed, and on its physical condition, such as hardness, temperature, etc. In the case of the water pipe, the length and diameter of the pipe, the nature of its walls, and the number of its bends, will determine a certain *hydraulic resistance*, which will permit the flow of water under a given head or pressure through it, and determine the amount which will escape from the reservoir in a given time. In the same manner, in an electric circuit, the length and cross-section of the conducting wire, or circuit, taken in connection with its nature and physical conditions, will determine a certain *electric resistance*, which will permit the flow of electricity through it, under a given pressure or E. M. F., and determine the amount of current which will flow through the circuit in any given case.

The law which determines the current strength in amperes, which will pass through any circuit under the influence

HEATING OF BARE CONDUCTORS. 39

of a given E. M. F. and against a given resistance in a circuit, was discovered by Dr. Ohm, of Berlin, and is known as *Ohm's law*. This law may be stated as follows: The current strength in any circuit is equal to the E. M. F. acting on that circuit, expressed in volts, divided by the resistance of that circuit, expressed in units of electrical resistance called *ohms*; or concisely, Ohm's law may be expressed as follows:

The amperes in any circuit equal the volts divided by the ohms.

For example, if a storage cell, with an E. M. F. of two volts, be connected to a circuit whose resistance, including that of the cell, is 10 ohms, the current strength passing through the circuit will be $\frac{2}{10} = \frac{1}{5}$ ampere; and, since one ampere is one coulomb per second, there would be flowing in such a circuit one-fifth of a coulomb per second. The work done in the circuit will be equal to the pressure of two volts multiplied by the total number of coulombs that pass in any given time. For example, in ten minutes, or in 600 seconds, the total number of coulombs that will have passed through the circuit will be $600 \times \frac{1}{5} = 120$ coulombs, and the work expended by the storage cell in the circuit will be $2 \times 120 = 240$ volt-coulomb, or joules, $= 177$ foot-pounds. We also know that the activity in this circuit will be the product of the volts and the amperes, or 2 volts $\times \frac{1}{5}$ ampere $= \frac{2}{5}$ watt $\frac{2}{5} =$ joule-per-second $= 0.295$ foot-pound-per-second.

During the flow of water through a pipe there will be produced a certain back pressure, or *counter-hydraulic pressure*, tending to check the flow of water through the pipe.

HEATING OF BARE CONDUCTORS.

In the same way, during the flow of an electric current through a conductor, there will be produced a *back electric pressure*, or *counter E. M. F.*, equal in all cases to the E. M. F. impressed upon the conductor. In fact, the current strength through the conductor adjusts itself in accordance with Ohm's law, in such a manner that the counter E. M. F. shall just be equal to the *impressed E. M. F.*; *i.e.*, the E. M. F. acting on the circuit. The counter E. M. F. in volts, is equal to the product of the current strength in amperes, by the resistance of the conductor in ohms. Thus, the 10-ohm circuit above referred to, carrying a current of one fifth of an ampere, develops a counter E. M. F. of $10 \times \frac{1}{5} = 2$ volts, which is just equal to the impressed E. M. F. of the cell. The product of the current strength and the counter E. M. F. is the activity expended in the circuit, just as the product of the current strength and the E.

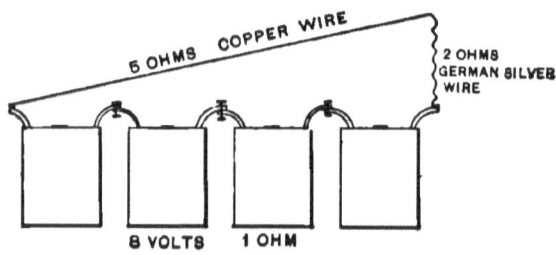

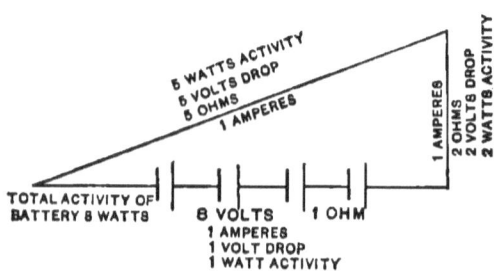

Fig. 1. —Distribution of C. E. M. F. in a Circuit.

M. F. is the work expended by the E. M. F.

If, for example, as in Fig. 1, four storage cells, each of 2 volts E. M. F. and ¼ ohm resistance, be connected in series with an external circuit composed of two parts; viz., of a resistance of 5 ohms of copper wire, and of a resistance of 2 ohms of German silver wire, the total resistance of the circuit will be 8 ohms, and the current strength $\frac{8}{8} = 1$ ampere. The back pressure, or *drop*, in the German silver wire will be $2 \times 1 = 2$ volts. The back pressure, or drop, in the copper wire, will be $5 \times 1 = 5$ volts, and the activity expended in each will be 2 volts × 1 ampere = 2 watts in the German silver and 5 volts × 1 ampere = 5 watts in the copper.

A counter E. M. F. may be produced not only by the back pressure of a current passing through a resistance, but also by the presence of certain devices placed in the circuit and operated by the current, such, for example, as electric mo-

44   ELECTRIC HEATING.

tors, or electrolytic cells. For example, if the circuit represented in Fig. 2 have

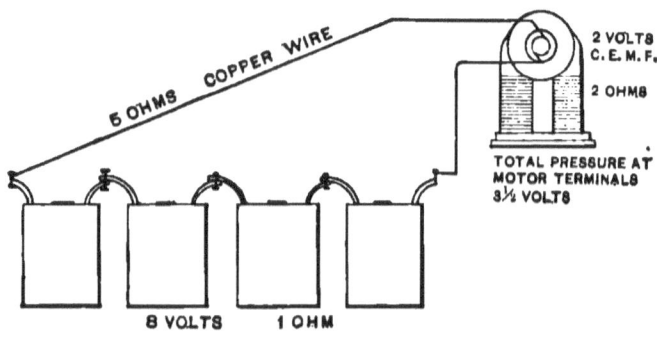

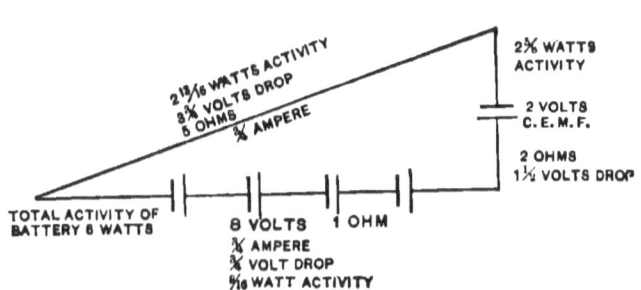

FIG. 2.—DISTRIBUTION OF C. E. M. F. IN A CIRCUIT.

its German silver wire of 2 ohms resistance, replaced by a small electromagnetic motor of 2 ohms resistance, and two volts counter E. M. F., this E. M. F. being developed by the rotation of its armature, then the current strength through the circuit will be 8 volts — 2 volts = 6 volts effective E. M. F. divided by 8 ohms resistance = $\frac{6}{8} = \frac{3}{4}$ ampere. The drop in the resistance of the motor would be $2 \times \frac{3}{4} = 1\frac{1}{2}$ volts, and the total C. E. M. F. of the motor $2 + 1\frac{1}{2} = 3\frac{1}{2}$ volts. The total work expended in the circuit by the storage cell will be $8 \times \frac{3}{4} = 6$ watts, and the total activity absorbed by the motor will be $3\frac{1}{2} \times \frac{3}{4} = 2\frac{5}{8}$ watts. Of this activity that part will be expended in heat which is developed in the resistance of the wire; namely, $1\frac{1}{2} \times \frac{3}{4} = 1\frac{1}{8}$ watts, and the remaining, or $1\frac{1}{2}$ watts, $= 2 \times \frac{3}{4} = 1\frac{1}{2}$ watts will, disregarding certain losses which occur in the revolving parts, be expended mechanically by the armature.

It will be noticed, in the above case, that the activity in the circuit, which is the product of current strength and counter E. M. F. due to resistance, is expended in heating the conductor, while the activity which is the product of current strength and counter E. M. F., due to what is called *magnetic induction*, is work expended magnetically. This may be taken as a general law; for, whenever a counter E. M. F. in a circuit is due to thermo-electric, chemical, or magnetic effects, the activity of the current on that C. E. M. F. is expended thermo-electrically, chemically, or magnetically; but when the C. E. M. F. is merely that due to the drop of pressure in the conductor, the activity in this drop is expended as thermal activity.

Consequently, when an electric source,

such as a dynamo-electric machine, is connected to a circuit, the counter E. M. F. of the external circuit must be equal to the pressure or E. M. F. of the dynamo at its terminals. The greater the proportion of this counter E. M. F. due to magnetic induction, or to chemical effect, the greater will be the activity expended in the circuit as magnetic, or as chemical activity, while the remainder, due to drop in pressure, or the resistance of the circuit, will be expended thermally in heating the conductor. When, therefore, a motor is connected to the terminals of a dynamo, the efficiency of the motor will increase with the proportion of the counter E. M. F. due to the rotation of the armature; whereas, if instead of obtaining mechanical work from the motor we wish to produce as much heat as possible in the circuit, we cause the motor to come to rest, so that all the electrical activity will be expended in the drop of pressure which will then constitute the entire counter E. M. F.

The resistance of any wire depends upon its *resistivity*, (or the resistance of a cubic centimetre measured between opposed faces) its length, and its area of cross-section (1 in. = 2.54 centimetres. 1 sq. in. = 6.4516 square centimetres. 1 cu. in.= 16.387 cubic centimetres.)

The following is a table of resistivities of the more important metals expressed in *microhms*, or millionths of an ohm, for a temperature of 0° C., the freezing point of water:

TABLE OF RESISTIVITIES.

Substance.	Resistivity.
Silver, annealed,	1.500 microhms.
Silver, hard drawn,	1.53 "

Copper, annealed,
(Matthiessen's
standard) .... 1.594   microhms.
Copper, hard drawn, 1.629   "
Iron, annealed, ... 9.687   "
Nickel, annealed, . 12.420   "
Mercury, liquid, .. 94.84   "
German silver, about 20.9   "

The reference to a standard temperature is necessary, in a table of resistivities, because the resistivity usually varies appreciably with variations in the temperature. Thus, the resistivity of pure soft copper is given as 1.594 microhms at 0° C. and this means that the resistance between any such pair of opposed faces as a and b, in a block of this copper one centimetre cube, as represented at A, in Fig. 3, would have a resistance of 1.594 microhms, or $\frac{1.594}{1,000,000}$ ohms.

If a wire having a cross-section of 1 sq. cm. as a^1, at B in Fig 3, have a length of 5 cms., then the resistance between the terminal faces a^1 and b^1, will be 5 times as great as between the terminal faces of the cube at A, in the same figure, or 5 ×

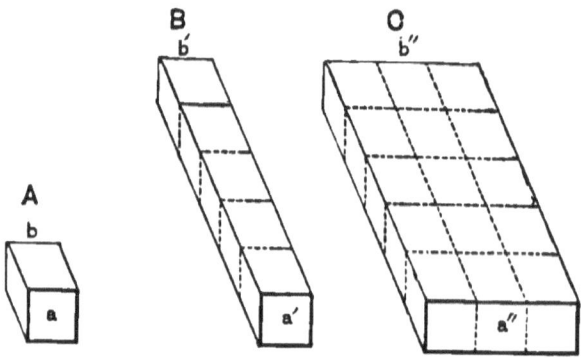

FIG. 3.—DIAGRAM REPRESENTING RELATION BETWEEN RESISTIVITY AND RESISTANCE.

1.945 = 7.97 microhms. Again, if the wire were 5 centimetres long, and had a crosssection of three square centimetres, as shown at C, in Fig. 3, then each centimetre length of such wire would have onethird the resistance of the unit cube, or

HEATING OF BARE CONDUCTORS. 51

$\frac{1.594}{3} = 0.533$ microhm, and the total resistance between the terminal faces a'' and b'', would be $0.533 \times 5 = 2.657$ microhms. In all cases, therefore, with a wire of uniform material, temperature and resistivity, it is only necessary to multiply the resistivity by the length in cms. and divide by the cross-sectional area of the wire in square centimetres, to obtain the total resistance of the wire.

While the preceding is a fundamental relation, yet, in practice, it is not always necessary to determine the cross-section of the wire in square centimetres, and its length in centimetres, in order to compute its resistance. In English-speaking countries it is customary to express the diameter of a wire in thousandths of an inch, or in *mils*, one mil being the one-thousandth of an inch. If we square the number of mils in the diameter of a wire we obtain the number of what is called *circular mils* in the wire. Thus, if a wire have a diameter of one-tenth of an inch = 100 mils, the number of circular mils in the cross-section of this wire will be 100 × 100 = 10,000 circular mils. A wire one inch in diameter would have a cross-section of one million circular mils.

The resistance of a pure standard copper wire one foot long, and one circular mil in cross-section, is 10.35 ohms, at 20° C., that is to say, a wire one-thousandth of an inch in diameter and one foot long would have this resistance. The resistance-per-foot in any pure copper wire will be this resistance, divided by the number of circular mils in its cross-section. For example, the wire above referred to as having 10,000 circular mils in its area of cross-section would have

HEATING OF BARE CONDUCTORS. 53

a resistance per-foot of $\frac{10.35}{10,000} = 0.001035$ ohm-per-foot at 20° C. The resistance of such a wire per mile would be $5280 \times 0.001035 = 5.465$ ohms.

While the use of circular mils is very convenient for wires whose length is expressed in feet, when tables or data concerning the resistance of a circular-mil-foot have been prepared, yet it is desirable to retain also the fundamental conception of the resistance as dependent upon resistivity and dimensions for the cases which may occur that are not dealt with in tables. For example, a resistance of 100 metres (10,000 cms.) of pure soft copper wire at 0° C. having a cross-section of 0.05 square centimetre would be $\frac{1.594 \times 10,000}{0.05}$ microhms = 318,800 microhms = 0.3188 ohm.

The resistivity of a metal is always reduced by the process of softening or annealing it, although the reduction in the resistivity, due to annealing, may only amount to one or two per cent. The resistivity depends very greatly, however, upon the physical nature and purity of the material. A very small percentage of certain impurities in a copper wire, such, for example, as phosphorus or sulphur, will greatly increase its resistivity, and even the presence of gases occluded or absorbed by the substance of the wire is said to appreciably increase its resistivity. The purity with which copper wires can be commercially obtained, at the present time, is such that their resistivity is, perhaps, only one per cent. greater than that of the so-called pure, standard, soft-copper wire, while it sometimes happens that wires are obtained commercially whose

HEATING OF BARE CONDUCTORS.

resistivity is actually less than that of this standard.

In dealing with wires of other metals than copper, such as lead, iron and German silver, the tabular resistivities cannot, as a rule, be relied upon to limits closer than say five per cent., and where a degree of accuracy greater than this is required, measurements of the resistivity of such wires, at a given temperature, are necessary. This can be done by carefully measuring the resistance of a given length of wire when its cross-section is known or can be carefully observed. The resistivity in ohms, at the temperature of the measurement, will then be the resistance multiplied by the cross-sectional area of the wire in square centimetres divided by the length of the wire in centimetres.

The effect of temperature on all pure metallic conductors is to increase the resistivity. Nearly all alloys also increase in their resistivity with increase in temperature, though less rapidly than their pure component metals. A few specially prepared alloys, such as platinoid, have a very small increase of resistivity with temperature, and are, therefore, in special request for the manufacture of permanent resistance coils, whose resistances are to remain as nearly constant as possible; while one or two alloys have been prepared whose resistivities are either not effected by temperature, or have a slight positive or negative coefficient; *i. e.*, a slight increase or decrease in resistivity with temperature, at different points of the thermometric scale. Carbon diminishes in resistivity about 0.5 per cent. per degree centigrade, reckoned from its

HEATING OF BARE CONDUCTORS.

resistivity at zero centigrade. Pure metals, or metals containing only a very small percentage of impurity, usually increase about 0.4 per cent. in their resistivity, per degree centigrade, above that which they possess at zero centigrade. For example, taking the resistivity of copper as 1.594 microhms at 0°C, its resistivity at 20° C. will be increased by 20 × 0.4=8 per cent., so that its resistivity at this temperature will be $1.594 \times \frac{108}{100} =$ 1.721 microhms, approximately. At the boiling point of water, or 100° C., its resistivity will have become increased by approximately 100 × 0.4 = 40 per cent., and its resistivity will be $\frac{1.594 \times 140}{100} = 2.232$ microhms.

When the resistivity of a wire is known, either by actual measurement at the temperature of observation, or from its tabular resistivity at 0 C. referred as above to the actual temperature, the amount of heat which will be developed in it in a given time, by a given current strength, becomes known, except in so far as its temperature elevation under the heating influence may be undetermined. For example, if a copper wire were insulated by a thin coating of some non-conducting varnish and placed in ice-water at $0°$ C., the resistivity of the wire might be 1.6 microhms, and a circular-mil-foot of this wire would have a resistance of 9.625 ohms. If the diameter of the wire were $0.01''$; i.e., No. 30 of the American wire gauge (A.W.G.) having a cross-section of 100.5 circular mils, the resistance of 10 feet of such wire would be $9\frac{6.25 \times 10}{100.5}$ $= 0.9577$ ohms at $0°$ C. If a current of two amperes be sent steadily through this length of wire, the drop in the wire would

HEATING OF BARE CONDUCTORS.

be 2 × 0.9577=1.9154 volts, and the activity expended thermally in the wire would be 2 × 1.9154=3.831 watts, or joules-per-second = 2.827 foot-pounds-per-second. The heat which would be expended in the wire would fail to appreciably raise its temperature, since it would readily pass through the insulating varnish into the ice-water, and, if we assume that abundant ice is present, the temperature of the water would not be raised until all the ice was melted. The work done by the electric source in supplying the current through this wire would, therefore, be expended in melting the ice.

If, however, the same length of wire be suspended in air, and the same current strength, of say 2 amperes, passes steadily through it as before, then, although some of the heat would be carried off by the air, yet the resistance offered by the air to the escape of the heat from the wire would be much greater than that offered by the varnish and water in the preceding case, so that the temperature of the wire would be raised. This would increase the resistivity of the wire at the rate of, approximately, 0.4 per cent. per degree centigrade of temperature elevation, so that both the resistance and the thermal activity of the wire would rise.

Suppose, for example, that the air surrounding the wire is at a temperature of 20° C. and that the current through the wire raises its temperature 10°C. above the surrounding air, or to 30° C. Then the resistivity of the wire before the current passed through it, would be $1.6 \times \frac{108}{100} = 1.728$ microhms, and after the current has passed through it steadily $1.6 \times \frac{112}{100} = 1.792$ microhms, so that the resistance of the heated wire will be 10.72 and the thermal activity in the heated wire 4.288 watts.

It is, therefore, a simple matter to determine the thermal activity in a given conductor when the drop of pressure in the conductor and the current strength passing through it are observed; for, if the drop in a wire, for example, be 5 volts, and the current through the wire, 100 amperes, then the thermal activity in the wire will be 500 watts. But it is by no means a simple matter to determine what temperature the wire will attain when subjected to this heating, since the wire is constantly losing its heat at a rate which depends upon a variety of circumstances.

When a current passes through a wire, the heat developed by that current causes it to increase its temperature. When a body is heated above the temperature of surrounding bodies, heat flows from the former to the latter, just as water flows from a higher to a lower level. The greater the elevation of temperature of the heated body, the more rapid will be the passage of heat, or the greater the *thermal current strength*. When the body is supplied with heat at a steady rate, its temperature continues to rise until the rate at which it receives heat is balanced by the rate at which it loses it. Consequently, a time is reached when the temperature of the body remains constant, although the body is constantly receiving heat. When, therefore, an electric current has been passing for a sufficient length of time through a conductor, its temperature will attain a definite elevation above that of surrounding bodies and remain constant, the thermal activity within the conductor being balanced by the loss of heat from the surface of the conductor.

Heat escapes from a body in three ways; namely,

(1) By *conduction* to bodies in immediate contact with its surface; as, for example, when a heated wire is enclosed in a mass of lead or rubber, the heat passing directly across the surface of the wire into the surrounding substance.

(2) By *convection*, which occurs only in fluids; *i.e.*, liquids or gases. Here, the particles of fluid surrounding the hot body become heated and are carried through the fluid mass by currents, set up by differences in density of the hotter and cooler portions of the fluid.

(3) By *radiation*, the heat passing out

from the heated surface in straight lines just as light does, when a body becomes incandescent.

As to which of the above methods of loss of heat is the most effective in the case of a wire heated by an electric current, depends upon the character of the surroundings of the wire, whether the wire is bare or covered, and where it is placed.

Circuit wires may be either bare or covered. Bare wires are only suitable for suspension in air. Covered wires may be placed in air, in water, or in the ground. The character of the covering may also vary in different cases.

It might be supposed that a bare wire suspended in the air was the simplest case to deal with. Such, however, is far from being the case; for not only does the position of the wire itself greatly affect the ease with which it loses heat, but also the condition of the surrounding air, whether at rest or in motion.

When a bare wire is supported horizontally in the air of a room, and an electric current is passed through it, this current will set up a certain drop of pressure in the wire, and the product of this drop and the current strength, will give the thermal activity developed in the wire at the outset.

Under these circumstances the temperature elevation of the wire will have become practically constant in about two minutes. As soon as this limiting temperature is reached the heat developed by the electric current in any length of the wire, such as an inch or a centimetre, will be equal to the heat dissipated from its surface by radiation and convection. The amount of heat that will be radiated in a given time, say one second, from a given length of the wire, say one inch, will depend upon the extent of free surface of the wire in that length, upon the nature of its surface, whether bright, blackened or colored, smooth or rough, etc., and upon the temperature elevation the wire has attained. A rough, blackened surface will radiate heat, approximately, twice as rapidly as a smooth, bright surface.

The heat which will escape from the wire by convection, in the same length, so far as is known, is practically the same for all diameters of wire and for all characters of surface, so that the loss by convected heat does not depend upon the surface, or only increases slightly with the surface, while the loss by radiated heat increases directly with the surface.

For every degree centigrade of temperature elevation attained by the wire above the surrounding still air of a room, the heat lost by convection is, approximately, 0.053 joules-per-second, per foot of length, so that if the wire has a temperature elevation of 20° C., every foot will lose by convection, approximately, 1.06 joules-per-second, or will lose heat energy at the rate of 1.06 watts. The loss by radiation will be approximately 0.004 watt per square inch of bright surface, per degree centigrade of temperature elevation.

The total loss of heat in watts will, therefore, be the temperature elevation of the wire, in degrees centigrade, multiplied by the number of feet, and by 0.053 for the effective loss and the same temperature elevation multiplied by the number of square inches of surface and 0.004 for the radiation loss.

When the air, in which a wire carrying an electric current is suspended, is in a state of motion, as, for example, when the wire is suspended out of doors, and exposed to wind and air currents, the loss of heat by convection from its surface is greatly increased even in the calmest weather. Air currents carry off a large amount of heat from the wire, so that the temperature elevation of the wire for a given current strength is considerably reduced.

CHAPTER IV.

ELECTRICAL HEATING OF COVERED CONDUCTORS.

An electric conductor, when employed to carry an electric current to a distance, is intended to be kept as cool as possible; first, because a hot wire necessarily means a wire in which energy that might otherwise be utilized is being expended uselessly as heat; second, because the resistance of a hot wire is higher than that of a cold wire and, consequently, more energy is wasted in the wire to sustain a given current; and third, because a wire that is overheated by the current it carries, may either destroy its insulation or set fire to inflammable bodies in its vicinity. On the contrary, an electric conductor, which is intended for purposes of developing heat by the expenditure of electric energy, as in an *electric heater*, is doing its best service when it is as hot as it can become without danger of injury from an excessive temperature. Since the great majority of heated electric conductors are those in which heat is both an objection and a loss, it is necessary to examine the laws which control their heating, with a view of avoiding a dangerously high temperature.

Whether a covered wire be supported in air, buried in the ground, or immersed in water, it is evident that its heat must first escape into the insulating covering, before it can pass into the surrounding medium. In other words, the insulating covering offers a certain resistance to the escape of heat from the wire, and, if

HEATING OF COVERED CONDUCTORS.

the covering could be removed without allowing the electricity to escape from the wire, the temperature of the wire, under any given current strength, would be less than that it attains with the covering in place.

The *thermal resistance* of any insulating covering, on a round wire, depends on the proportion of the diameter of the bare conductor to the diameter of the covered conductor, and on the nature of the insulating material. As no two insulating coverings offer exactly the same electric resistance to the escape of electricity, so no two insulating coverings offer exactly the same thermal resistance to the escape of heat from the wire. All good electric insulators are good thermal non-conductors, so that just as a considerable difference of electric pressure is required to force a given quantity of electricity through a conducting coating on a wire, so a considerable *difference of thermal pressure;* i.e., difference of temperature, is required between the inside and outside of the coating to force a given quantity of heat through the coating. When, therefore, the insulating coating is thick, it is to be expected that the temperature elevation of the wire, for a moderate current strength, will be appreciable. If, however, the covered wire be supported in the air of a room, it will frequently happen that the wire will be cooler than if devoid of covering, for the reason that the advantage gained by increased external surface and the greater radiation therefrom, will more than compensate for the additional thermal resistance between the surfaces of the wire and the air surrounding it. The same is also more likely to be the case if the insulating covering of the wire be blackened, since its radiation will thereby be increased.

When a covered wire, instead of being supported in air, is immersed in water, the temperature elevation of the wire is increased by reason of the insulating covering; for, if the wire could be covered with a very thin, electrically non-conducting varnish, it would be almost impossible to raise the temperature of the conductor, so rapid is the communication of heat from the metal to the mass of surrounding liquid, and so slow the elevation of temperature in the liquid, if its volume is large. With air, as we have seen, the case is very different; the thermal resistance of still air is often large, while the thermal resistance of water is very small. With wires submerged in water it may be safely assumed that the entire thermal resistance to the escape of heat exists in the non-conducting covering, and that no thermal resistance exists in the water outside it.

A covering of metal on the external surface of an insulated wire, such, for example, as a thin shell of lead spread over the insulating material, does not offer any appreciable thermal resistance. Metals conduct heat so rapidly, as compared with insulating substances, that the thermal resistance in the metal may be neglected. In fact a lead sheath aids in cooling a wire suspended in air, since it provides an increased surface for loss of heat by radiation and convection; or, in other words, it reduces the effective thermal resistance of the air.

The *safe carrying capacity* of a conductor may be defined as the current strength that can safely be permitted to pass through it. The carrying capacity depends upon the highest limit of temperature elevation permitted as consistent with safety. In some cases, it is desirable, from considerations of economy of installation, to press the electric activity of a wire up to the limit of safety. In most cases, however, it is too expensive to force the activity of a wire to such a limit, for the reason that the expense of the thermal activity expended in the wire, at the safety limit, renders a larger and more costly wire, with a lower resistance and diminished temperature elevation, economical. In cases where it is desirable to carry a powerful current with the minimum cross-section or weight of conductor consistent with safety, it is often advantageous to subdivide the conductor; *i. e.*, to employ two or more small wires instead of a large single conductor. In the case of a subdivided conductor, the temperature elevation of each separate wire will be considerably less than the temperature elevation of a single wire carrying the entire current. This is for the reason that the surface of a pound of a given wire varies with its area of cross-section, decreasing as the area of cross-section increases, and vice versa. In other words, a small wire has a larger surface, per pound, than a large one, and, as is evident, the greater the surface, the greater the rapidity with which the heat generated within the substance of the wire can escape.

An insulated wire placed in a wooden moulding, or in a closely-fitting conduit in

HEATING OF COVERED CONDUCTORS.

a building, loses its heat entirely by conduction, provided the walls of the panel or conduit are everywhere in contact with the external surface of the covered wire. In this case, the temperature elevation of the wire, for a given current, is greater than if the wire were immersed in water, since the thermal resistance of the walls of the panel is added to the thermal resistance of the insulating covering. In almost all cases, however, the temperature elevation is less than if the wire were supported in air. Consequently, the effective thermal resistance of a panel or conduit, is generally less than the effective thermal resistance of the air within a room.

The rule in common use for determining the size of wires to be placed in wooden mouldings, is to allow 1000 amperes per square inch of area of cross-section. This rule is easily applied, and affords a convenient guide in the absence of any special tables of reference. It must be remembered, however, that the rule implies that a large wire will lose its heat as readily as a small one, and this, as we have seen, is not the case, owing to the reduction of surface per unit of cross-sectional area or weight. Consequently, a very large wire, selected according to this rule, would be heated to a much higher temperature than a very small wire. In fact, the rule is not to be regarded as entirely safe beyond 250 amperes of current strength.

In buildings which are not absolutely fireproof, it is important that the conductors, which may be placed in them for supplying electric light or power, shall be so proportioned that their temperature may never become dangerously high. A wire which can be grasped in the hand, say for a minute, without marked discomfort from its heat, may be regarded as at a safe temperature. The limiting temperature, defined in this way, will of course depend physiologically upon the condition of the hand and the sensibility of the person making the experiment, but roughly, may be considered as in the neighborhood of 50° C. If we assume that the summer temperature of the interior of a house is 30° C. or 86° F., then to conform with these requirements as to temperature, the limiting temperature elevation for such a wire would be fixed as approximately 20° C. In other words, we must not allow the current strength through the wire to exceed that necessary to elevate its temperature 20° C., since, otherwise, in summer, the temperature attained by the wire at full load would exceed 50° C. In practice, however, the limiting temperature allowed by Fire Insurance Boards is sometimes placed as low as 10° C. at full load, so as to allow margin for any accidental overloads that may occur unexpectedly.

If we double the current strength passing through a wire, under any given conditions, we quadruple, roughly, the temperature elevation of the wire. Thus, if a wire in moulding be elevated 10° C. above surrounding temperatures by the passage of its full-load current, then twice that current strength will elevate its temperature 40° C., approximately, or 72° F., and if the wire be originally at a temperature of 78° F., its final temperature with double full load will be 150° F.

Insulated wire for underground work usually possesses in addition to the ordinary insulating material, a sheathing of lead, and is either buried directly in the ground, or is placed in a conduit. The necessity for obtaining a ready access to wires for their examination has led to the latter process being adopted in most cases. In order to insure high insulation, the conduits frequently have air forced through them, in which case their condition approximates to that of a lead-covered wire supported in air.

Taking now the case of a wire buried directly in the ground, the thermal resistance to the escape of heat from the conductor is not only that of the insulator, but also that of the ground. If the ground be moist, its effective thermal resistance is reduced, but if it be dry, the thermal resistance may be considerable. In almost all cases, however, the thermal resistance of the ground is less than the thermal resistance of still air, so that a buried wire, carrying a given current strength, will be cooler than the same wire supported in still air, although cases may occur in which this statement does not hold good.

Intermediate between the condition of a wire suspended in the air of a room, and a wire in a conduit, in which there is no attempt at forced ventilation, there is the condition of a wire supported in a subway. Here the air being at rest, the conditions approximate, thermally at least, to the case of a wire in the still air of a room.

When a wire has been electrically inactive for a considerable period of time, its temperature will necessarily coincide with that of the surrounding air or other material. When, however, the full-load current is sent through the wire, its temperature will immediately begin to rise, the rate of elevation of temperature being a maximum at the outset, and diminishing steadily as elevation of temperature is attained. From a theoretical standpoint the wire never does reach the full maximum temperature, but always approaches it. Practically, however, a wire in air, reaches, say 95 per cent. of its maximum temperature in two minutes after the application of the full-load current strength. In water a wire reaches this temperature in about ten minutes after the full-load current is applied; in wooden moulding, in about fifteen minutes, and, when buried in the ground, in about twenty minutes. The larger the wire, the greater will be its mass, and, consequently, the longer the time required by it to attain its full temperature elevation.

In the case of buried wires, the heat has to be propagated slowly outward from the wire through the mass of the neighboring earth. The result is that, while the layers of earth closely surrounding the wire will probably reach 95 per cent. of their maximum temperature elevation in half an hour, the layers situated at a considerable distance from the wire, although they will necessarily receive a much smaller temperature elevation, yet will require a much longer time for that temperature elevation to be established, and many hours may elapse before 50 per cent. of the maximum temperature elevation is attained at a distance of say four feet from a deeply buried wire.

The temperature elevation, which may be permitted in a wire buried in the ground, is determined by totally different conditions to those which limit the temperature elevation of a wire placed in a building; for it is evident that there is no danger of setting fire to the ground. The insulating material of a wire has, however, to be sufficiently plastic to allow the wire to be bent or slightly stretched, and this condition, together with good electric insulation, is usually found in a substance that will not permit of a high temperature without injury. Even if it were possible to operate a buried conductor at a high temperature, such temperature would be dangerous where the conductor emerged from the ground. The temperature elevation, in the case of hemp-covered wires, is usually 25° C. and in rubber-covered wires 20° C. Most insulating materials, long before they would be injured by the heat, would be liable to soften, thus permitting the conductor to sag, so that it would no longer remain embedded centrally in the insulating material. Consequently, the permissible temperature elevation is limited by the softening point.

As regards the temperature elevation of ocean cables, employed in submarine telegraphy, the question is at present devoid of practical interest, since the currents which such cables carry are so very feeble, say generally only a few milli-amperes, that the temperature elevation of the conductor is entirely negligible. It is worth pointing out, however, as an interesting fact, that should occasion ever arise for sending powerful currents through submarine cables, the fact that the entire bed of the deep ocean is covered by a layer of very cold water in the neighborhood of 30° F., would permit a ready loss of heat.

CHAPTER V.

FUSE WIRES.

A WIRE placed in a building, although so proportioned relatively to the current strength it has to carry, that, under ordinary circumstances its temperature will be perfectly safe, yet, owing to accidental external causes, the current strength may sometimes become enormously increased, thereby heating the wire to a dangerously high temperature. If, for example, the wire has in its circuit a group of lamps, requiring normally 10 amperes of current from a pressure of 115 volts, then, if by some accident a short-circuit be effected at the lamps, the current strength through the lamps would be much diminished, but the strength of current in the wire, supplying the lamps, might become enormously increased; for, while the pressure on the mains would remain practically the same, the resistance in the circuit, if very small, would permit, by Ohm's law, a very powerful current to pass through it.

The effect of such an abnormally great current would be to cause the amount of heat liberated in the wire, forming the short circuit, to be far greater than it could dissipate without attaining a temperature sufficiently high to make it red hot, or even to melt it. If such a wire were melted by an accidental short-circuit, not only would there be danger of setting fire to the wood-work, or other inflammable material surrounding the wire, but there might also be considerable trouble and difficulty in replacing the wire after the accident. Moreover, the effect of a vio-

FUSE WIRES

lent overload, sometimes sufficiently great to melt even a stout conductor forming some portion of the circuit, would be liable to injure the dynamo or engine driving it, or to overheat and consequently injure any electrical apparatus that might be in the same circuit. In order to avoid these difficulties the plan has been universally adopted of inserting wires, called *fuse wires*, in the branch and main circuits of any system supplied by a dynamo.

A fuse wire is a wire or a strip of metal, which has both a high electric resistance per unit of length, and a low melting point. If such a wire be in circuit with a copper wire, and both are of such sizes that they are able to carry the normal, full-load current without overheating, it will be evident that the fuse wire must become much hotter than the copper wire;

for, since, as we have seen, the amount of heat developed in any circuit, the current strength remaining the same, depends on the resistance of the circuit, it is evident that the same quantity of heat will be developed in such lengths of the fuse wire and the copper wire, as have an equal *drop; i.e.*, offer an equal resistance to the current. Consequently, there will be developed in, say one inch of fuse wire, the same amount of heat as would be liberated in, perhaps, ten feet of copper wire. The fuse wire will, therefore, be raised to the temperature at which it melts, long before the temperature of the copper wire would pass the danger point, and the melting of the fuse wire would interrupt the circuit and thus automatically cut off the current. The meaning of the term *safety fuse* is, therefore, evident, since the simple introduction of such a wire into the circuit would absolutely prevent the passage through such circuit of a current that would raise its temperature to a dangerously high degree. It is fortunate that so simple a plan as the mere insertion of a safety fuse should be capable of protecting electric conductors against the consequences of accidental short circuits. Like many other inventions, its value lies largely in its extreme simplicity, and in the certainty with which it can be relied upon to operate effectively.

Fuse wires are composed of lead and tin, or tin-lead alloy. These wires usually occur in the sizes shown in Fig. 4. Here, on the right hand, the diameters of the wires are given in circular mils, and on the left hand, the carrying capacity of the wires in amperes. It is to be observed, that although the cross-section of a wire is quadrupled when its diameter is doubled, yet the carrying capacity is not

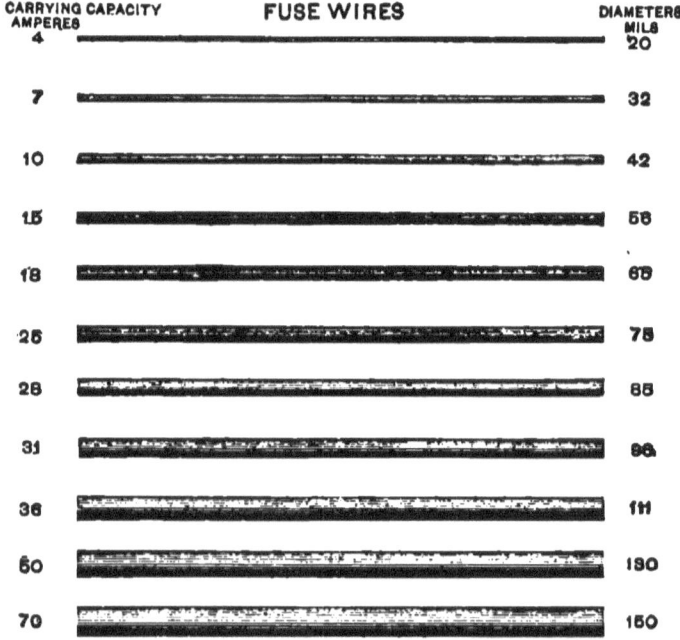

FIG. 4.—DIAMETER AND CARRYING CAPACITIES OF FUSE WIRES.

quadrupled. The carrying capacity increases faster than the diameter of the wire, but less rapidly than its area of cross-section.

FUSE WIRES. 93

Safety fuses are not only employed in the form of wires, but also in the form of strips, as shown in Figs. 5 and 6. In Fig. 5, the *safety strips* are connected to the circuit by means of binding posts, the studs of which pass through holes at each

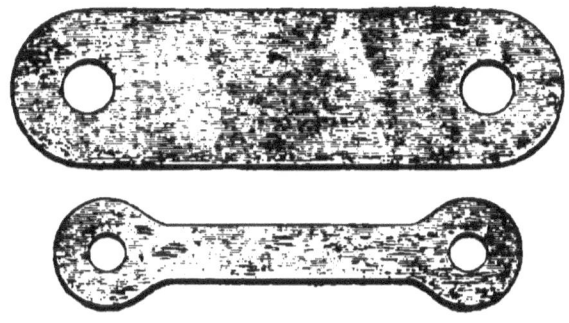

FIG. 5.—FUSE STRIPS.

end. In Fig. 6, the ends of the strips are slipped beneath the screw clamps, thus avoiding the necessity for the removal of the screw head, as would be the case in the form shown in Fig. 5.

Fuse wires, such as shown in Fig. 4,

94   ELECTRIC HEATING.

are placed in the circuit by simply wrapping them around binding posts connected with the circuit and firmly clamping the connection with a screw head. This pressure is apt to damage the wire and alter

Fig. 6.—Fuse Links.

its carrying capacity, thus causing it to melt at a unduly low strength of current. To avoid this, the ends of the wire or

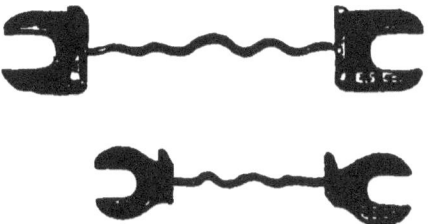

Fig. 7.—Copper-Tipped Fuse Wires.

strip are often fused into copper clamps as shown in Figs. 7 and 8. Large safety strips are usually of the form shown in Fig. 8, the lead strip being riveted to the copper end pieces.

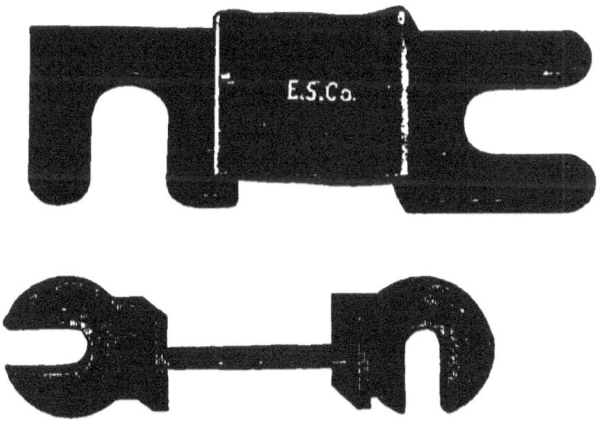

Fig. 8.—Copper-Tipped Safety Fuses.

96 ELECTRIC HEATING.

Fig. 9 shows a simple form of *safety fuse-block* consisting of a slab of slate, or other non-inflammable material, on which are mounted two metal blocks B and B. The circuit passes through these metallic blocks, and the fuse wire is clamped between them as shown.

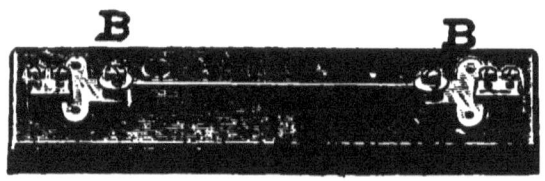

FIG. 9.—SAFETY FUSE-BLOCK.

Fig. 10 shows a pair of strip safety fuses S_1, S_2, or *safety links*, as they are sometimes called, inserted in the circuit of the two leads BB^1 and AA^1, under thumb screw clamps situated at the ends of the metallic blocks which form the terminals of the leads BB^1 and AA^1. These blocks are mounted on a non-conducting and non-inflammable plate, such as a slab of slate, porcelain, or marble.

Fig. 11 represents a *porcelain fuse-block* prepared for the reception of safety links between the screw clamps A, A^1.

FIG. 10.—PAIR OF SAFETY STRIPS, MOUNTED ON FUSE-BLOCK.

and B, B^1. The two supply mains A and B are electrically separated from each other by the porcelain projecting ridge RR, provided for this purpose. The pressure between these leads may be 100 or 200 volts, according to circumstances, and were the ridge not present, the blowing of the fuse might establish a dangerous

FIG. 11.—PORCELAIN FUSE-BOX.

arc across the leads, or such arc might be accidentally established during the proc-

FUSE WIRES.

ess of connecting the safety links and thus, perhaps, injure the attendant.

Fuse-boxes are generally provided with a porcelain cover, though at times, for the purpose of ready inspection, a transparent cover, such as glass or transparent mica, is employed. Figs. 12 and 13 show examples of fuse-blocks of the latter type with the fuse wires or links in position. The arrangement of the box will necessarily vary according to whether the main wires terminate in the box, or pass through it. Thus at A, Fig. 12, the mains pass directly through the box in the grooves on the left hand, but after being bared of their insulation, have their conductors clamped underneath the screws whose heads are visible in the grooves. Connections exist beneath the box from these screws to the safety links on the right-hand side and the branch wires are

100   ELECTRIC HEATING.

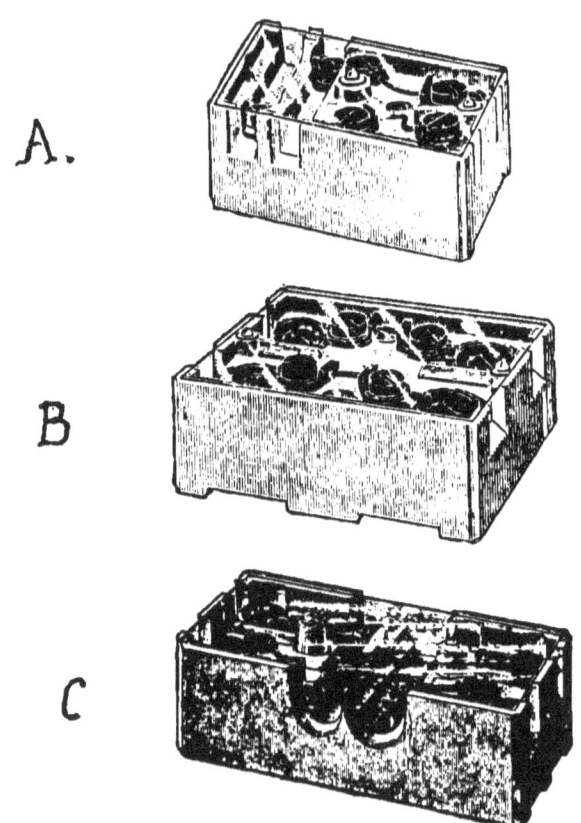

FIG. 12.—MICA-COVERED FUSE-BOXES.

carried off at right angles. In the event of any short-circuit between the branch wires, one or both of the safety links is melted, but no accident in the main circuit can affect these fuses, since the main

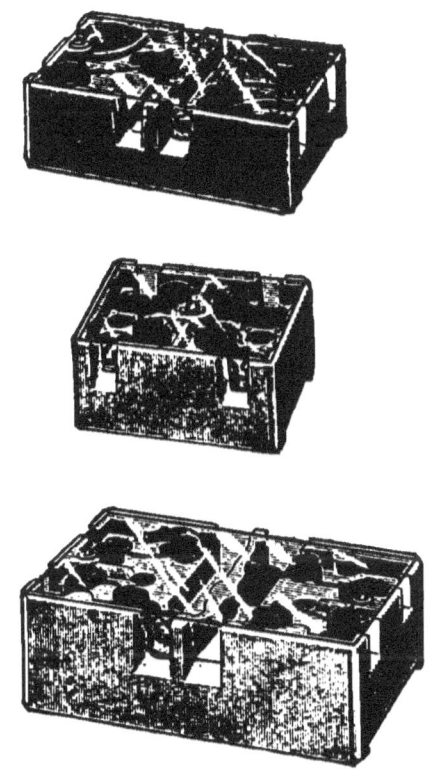

Fig. 13.—Mica-Covered Fuse-Boxes.

conductors, as already mentioned, pass directly through the box.

At *B*, is shown a form of safety fuse-box through which the mains do not pass, but terminate, say at the left, and the wires supplied by such mains enter at the right.

At *C*, a form is shown from which two separate branch circuits issue from the

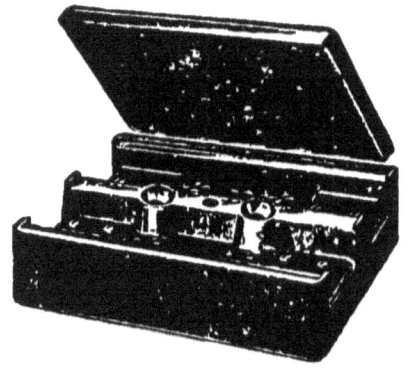

FIG. 14. —FUSE-BOX PROVIDED WITH PORCELAIN COVER.

box, half to the right and half to the left, after being suitably connected to the mains which enter and pass through the centre of the box.

Practically similar forms are shown in Fig. 13.

FUSE WIRES.

Fig. 15.—Fuse-Box with Fuses in Cover.

104   ELECTRIC HEATING.

In all these forms, a thin mica cover serves to exclude dust, and, at the same time, renders the conditions of the safety links externally visible.

Figs. 14 and 15 show forms of fuse-boxes, provided with porcelain covers.

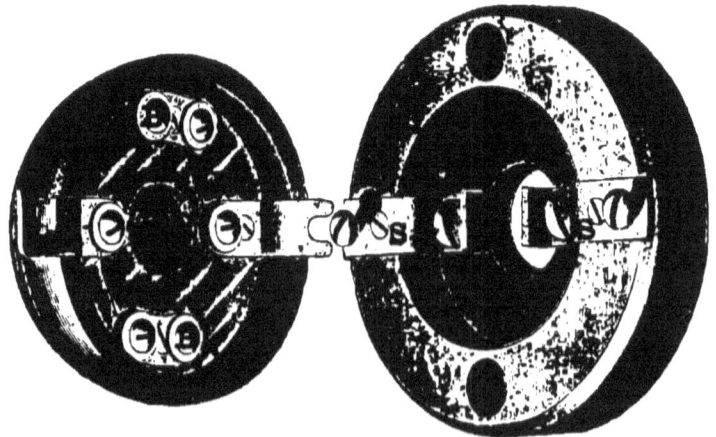

FIG. 16.—CEILING-FIXTURE FUSE-BLOCK.

The form shown in Fig. 14 is similar to the box shown in Fig. 11, with the addition of sides and cover. Fig. 15 shows a form of box in which the safety links are supported on the cover, and the wires con-

FUSE WIRES.

nected to the base, so that the attachment of the cover to the base closes the circuit through the links.

The form of fuse-box necessarily varies with the current which has to be carried through it, and with the character of the

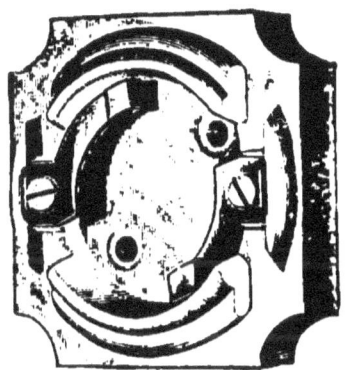 

FIG. 17.—CEILING BLOCK WITH SPRING CLIPS.

fixture or circuit in which it is connected. Fig. 16 shows a form suitable for a ceiling fixture; *i. e.*, an electrolier pendant from a ceiling and usually called a *ceiling block*. The supply wires are connected to the screws S, S, in the permanent block which is attached to the ceiling, while the wires connected to the electrolier are connected to the screws B, B, in the cover. Connection is secured through the two safety fuses F, F, by screwing up the cover against the block. A similar form is shown in Fig. 17, in which, however, connection is secured through spring clips.

Fig. 18.—Plug Cut-Out.

The fuse wire is sometimes placed in a screw-socket in order to ensure ease in placing and replacing. Under these circumstances the electrical connections of the fuse wire are such that the mere insertion of the screw block in its socket

FUSE WIRES.

inserts the fuse in the circuit. Fig. 18 shows such a *screw-block fuse*, or *plug cut-out* and Fig. 19 shows various forms of *socket attachments*, or *cut-out boxes*, for such fuses. The cavities of the block containing the fuse wires are usually part-

FIG. 19.—CUT-OUT BOXES.

ly filled with plaster-of-Paris for the purpose of excluding the air; for, when a fuse wire suddenly melts or blows, the heated air might escape explosively from the cavity forcing particles of melted lead outward. The effect of the plaster-of-

Paris on the action of the fuse, is to diminish its sensitiveness to a momentary overload, for the plaster conducts heat from the wire, and, therefore, a sudden excess of heat will not so quickly bring the wire to the melting point, although a steadily continued current will eventually melt the fuse almost as readily as if the plaster were absent.

When fuse-blocks are placed inside apparatus, it becomes a matter of importance to insure convenience in inserting and inspecting them, and when such apparatus, as, for example, an alternating-current transformer, employs dangerously high pressures, some means are necessary in order to insure safety of attaching the fuse wires to the fuse-block by disconnecting them from the primary and secondary terminals. A form of such a

FUSE WIRES. 109

fuse-block is shown in Fig. 20. Here an iron box *BB*, encloses a porcelain fuse-

FIG. 20.—TRANSFORMER SAFETY FUSE-BOX

box, whose cover *C*, is removed to show the interior. In this case, the porcelain fuse-blocks are detachable. One of them is shown at F, detached, and the other at F^1, in place of the interior. The fuse wire w, w, is clamped under screws whose studs project through the fuse-block and enter into spring clips p, p^1, when the fuse-block is thrown into position by its handle h, is connected with the external circuit by a wire shown, and P^1, connected to the apparatus in the interior. Should any short circuit exist in the apparatus, the fuse will melt on the insertion of the block, and the hand of the operator will be protected from any particles of exploded lead by reason of the shielding action of the handle h.

The temperature at which a fuse wire will melt, depends upon its composition. Some alloys can be used which will melt at as low a temperature as 50° C. As a rule, however, the melting point is about 300° C.

The current strength which will melt a fuse depends upon a variety of circumstances. It might be supposed that for a given diameter of fuse wire, the length of the wire forming the fuse would not influence its melting point. Such, however, is not the case. A long fuse wire will usually melt at a lower current strength than a short fuse wire, principally for the reason that the heat generated in a short wire is conducted by the metal in the wire to the metallic masses forming the clamps at each end, thus enabling the heat in the wire to be dissipated more rapidly than would be possible in the case of a longer fuse. Similarly, the position of a fuse wire, whether closely surrounded in a practically air-tight chamber or freely exposed to such currents of air as might exist in its vicinity, would greatly effect the current strength that melts it. So also the position of the wire, whether vertical or horizontal, its shape, whether straight or curved, the shape of its cross-section, the character of its surface, whether rough or smooth, tarnished or bright, all exert an influence on its carrying capacity. As a rule, therefore, fuses cannot be depended upon to melt at precisely the current strength for which they are designed.

When an overload, or an unduly powerful current, exists in an electric circuit for a very brief interval of time, as, for example, when a short circuit occurs during a small fraction of a second, a fuse designed to melt at say, 10 amperes, may carry 100 amperes or more without melting, when 10 amperes steadily maintained

FUSE WIRES. 113

for one minute would insure the melting of the fuse. This is for the reason that heat has to be expended in the mass of the fuse before its temperature can be raised to the melting point. Consequently, an appreciable fraction of a second may be required for even a powerful current to develop this heat; while, when 10 amperes flow steadily through it, ample time is afforded to bring up the temperature of the metal.

It sometimes occasions surprise that when a dynamo supplies a distant branch circuit through two fuses, one of which, a large fuse near the dynamo, called *the main circuit fuse*, is capable of carrying, say 500 amperes, and the other, a small *branch fuse* in a branch circuit, is capable of carrying only 20 amperes, that on an accidental short-circuit in the branch circuit, the main fuse should blow out, while the branch fuse remains intact. This action, by no means of common occurrence, probably finds its explanation in the fact that the main fuse has already been heated by a full-load current of the generator, to a comparatively high temperature, while the particular branch fuse is cold since no current had been passing through it prior to the accidental short circuit. Under these circumstances, when a short-circuit suddenly occurs between the branch wires, the powerful rush of current through both fuses may be able to blow the larger fuse, before the smaller one reaches the temperature of its melting point.

Since in most commercial electric circuits fairly considerable variations in the strength of the current passing are apt to exist without constituting either a dangerous or objectionable overload, if the carrying capacity of the fuses is made too near their normal-load current, considerable inconvenience may arise from the frequency with which the fuses are blown. For this reason, in good practice, fuses are generally employed whose carrying capacity is about fifty per cent. greater than the full-load current.

In central stations supplying underground systems of conducting mains, the inconvenience above pointed out arising from the blowing of fuses is so marked that in many cases such fuses are omitted entirely in the central station, and are only inserted between the mains and the consumers, as well as in all the branch circuits of the house wirings. Should, for example, a large feeder either become overloaded, or develop a short circuit at some point underground, it would probably blow its fuse, and the extra load would, therefore, be transferred to other feeders. These in their turn would also be liable to blow their fuses, until, in some cases, the entire system of feeders and mains might thus be cut off from the dynamos.

CHAPTER VI.

ELECTRIC HEATERS.

ONE of the commercial uses to which electricity has lately been applied has been the artificial heating of air in buildings on a comparatively small scale. While this method of obtaining artificial warmth has not yet reached such economy as to permit it to be economically applied to the heating of the air of large buildings, yet the convenience arising from the facility with which the electric current can be led to the electric heater, the comparatively small size and portability of the latter, the readiness with which the current can be turned on and off, the safety of the apparatus, its freedom from fumes or dirt, and the ease with which it can be managed, have attracted no little attention, and its use, in certain directions, is rapidly increasing. While there is, perhaps, little probability in the near future of large electric plants being erected whose current shall be entirely employed for the production of heat, as in warming buildings, nevertheless, electric heaters are likely to be extensively employed in connection with already existing systems of electric distribution for light and power.

Electric heaters are to-day in common use in electric street railway cars, and this is for the same reason that electric lights are employed in these cars. Were it not for the fact that the cars obtain their propelling power from the electric current, it is not at all likely that electrically lighted and electrically heated cars would have come into the general use they have to-day; although in parlor cars on steam railroads, electric incandescent lamps are sometimes employed as luxuries.

Electric heaters, designed for the artificial warming of air, though made in a great variety of forms, consist essentially of a metallic conducting wire, generally of galvanized iron, or German silver, loosely coiled so as to possess a comparatively extended radiating surface, and commonly supported in the air.

In order to obtain a sufficiently extended surface for radiation and convection, and also to obtain the desired electric resistance in the coil, within a limited space, it is usual to wind the wire in a loose spiral around a form or block of earthenware, porcelain, or other similar, non-inflammable material.

We have seen that a definite relation

exists between a given amount of electric energy and the heat energy it is capable of producing. It has been ascertained that one joule of work, expended in producing heat, will raise the temperature of a cubic foot of air about $\frac{1}{18}°$ F., and, therefore, an activity of one joule-per-second, or one watt, can raise the temperature of one cubic foot of air $\frac{1}{18}°$ F. per second.

A simple form of cylindrical electric heater for hot air is shown in Fig. 21. It consists of a metallic strip, wound spirally on an insulated frame. Here, as in all forms of air heater, the design is to obtain as large a surface exposed to the air as possible. Since the metal strip employed is comparatively thin, the total mass or weight of the metal in the heater is comparatively small, and the conductor is rapidly heated by the passage of the cur-

ELECTRIC HEATERS.

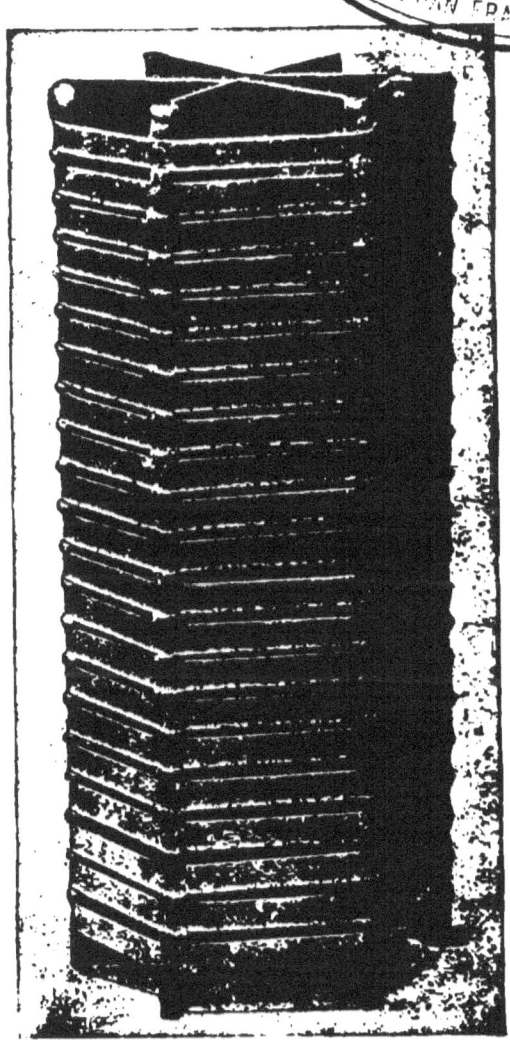

Fig. 21.—Cylindrical Electric Heater.

rent. But since the surface exposed to the air is great, the heating coil never acquires an excessively high temperature. An electric heating coil best serves its purpose when it rapidly imparts its heat to the surrounding air, never itself acquiring a dangerously high temperature.

The heating coil or conductor in an electric heater is not always in the form of a strip. It sometimes takes the form of a wire or spiral, either bare, or placed within a metallic frame.

Fig. 22 represents a form of electric heater or radiator resembling in appearance an ordinary steam or hot water radiator. Here the coils of the electric conductor are placed within the metallic frame. The exact length and dimensions of the heater coils will depend upon the amount of heat required, and on the electric pressure employed in the building. The same coil will, however, give practically the same amount of heat when con-

FIG. 22.—ELECTRIC RADIATOR.

nected with the same pressure of either alternating or continuous current.

The advantages of an electric heater are especially marked when employed in cars propelled by electricity. Indeed, the necessity for utilizing all the available space in a street car for the accommodation of passengers, and for maintaining a uniform temperature, with a minimum of attention required from the conductor of the car, renders the use of the electric current for heating even more economical than the use of a stove. This, of course, arises largely from the fact that the stove which can, in practice, be placed in the limited space allotted to it in a car, must necessarily be very uneconomical, moreover, the large scale on which electric power is generated in a central station for propelling the cars, reduces the cost of the electric energy so much that the electric heating of the car actually compares very favorably in economy with what would be required to heat it as effectively by the direct burning of coal in a stove.

Fig. 23 represents a form of *electric car*

heater, in front elevation, and Fig. 24, the back and interior of the same heater, showing the electric coil in position. Four or six of these heaters are employed in each car, according to the size of the car and the climate of the locality in which it

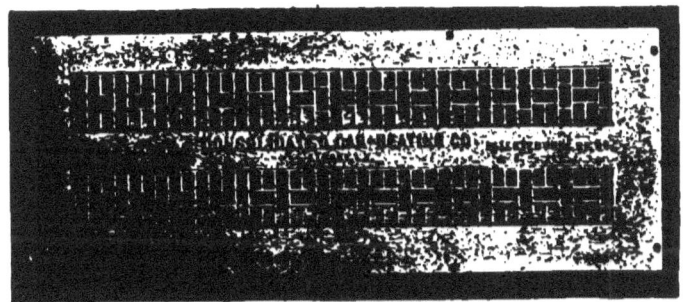

FIG. 23.—ELECTRIC CAR-HEATER.

runs. The heater is placed in a hole or gap made in the riser, or vertical partition, below the car seat. A cast-iron plate, furnished with grid openings, placed in the front of the heater and opening into the car, serves the double purpose of preventing the dress of the passengers from coming into contact with the heated coils, and for permitting the ready escape of the air through the apparatus.

An inspection of Fig. 24 will show that the heating coil, employed in this particular form of car heater, consists of a close

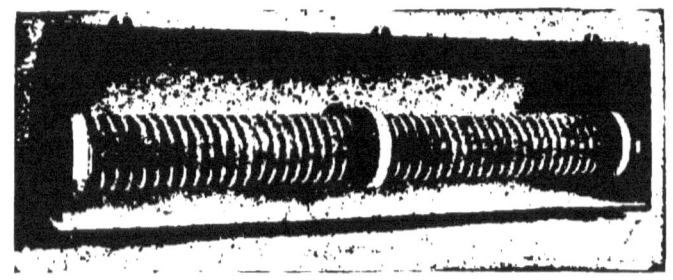

Fig. 24.—Back and Interior of Electric Car-Heater.

spiral conductor, which is spirally wound around a grooved porcelain tube, and is supported at the centre and at the two ends by porcelain washers. The back of the heater is formed of sheet iron, suitably provided with asbestos lining.

ELECTRIC HEATERS. 127

Heaters employed on electric railroad circuits take their current from the mains at a constant pressure, generally 500 volts. In order to vary the current passing

FIG. 25. –CAR-HEATER REGULATING SWITCH.

through the four or six heaters generally employed in each car, a switch is used, by means of which the separate heater coils can be connected in series, or in parallel-series, or some of them cut out from the circuit, thus permitting the amount of heat to be readily varied in order to meet the requirements of the

Fig. 26.—Side Interior View of Car-Heater Regulating Switch.

weather. Fig. 25 shows a form of *regulating switch* of this character intended to produce five different strengths of current, and, therefore, five different rates of producing heat in the car. The side view of the interior of the switch is shown in Fig. 26; the front view of the interior of the switch in Fig. 27. This switch consists of a number of contact springs,

Fig. 27.—Front Interior View of Car-Heating Regulating Switch.

whereby, through the motion of a lever attached to the barrel, the proper connections can be made for coupling the coils in the five different arrangements required.

The connections from the switch to the trolley wire and the ground through the various heaters, is shown in Fig. 28. In position No. 1 all the coils are connected in series, so that the current has to pass through each in succession. This position

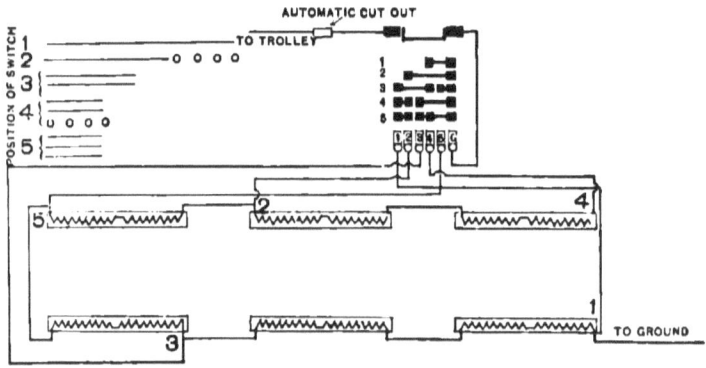

Fig. 28.—Diagram of Wiring for Six Electric Heater Equipment.

corresponds to the minimum current strength, about 2 amperes, and, therefore, to the minimum thermal activity, or rate of developing heat; namely, about one kilowatt. In position 2, two heaters are entirely cut out of the circuit, so that the resistance of the series being diminished, the current strength and activity in the remainder are increased, and the four active heaters will supply more heat to the car than the six heaters in the first case, the current being nearly 3 amperes, and the activity nearly 1500 watts. In the third position, the six heaters are connected in two series of 3 each, so that the current strength in each series is about twice that in the first position, or about $3\frac{1}{2}$ amperes in each series; *i. e.*, 7 amperes or 3.5 KW. in the combination. The fourth position connects two sets of two heaters and cuts out two heaters entirely. This gives about 4 amperes in each series, or 8 in the combination, representing 4 KW. In the fifth position, three rows of two heaters are employed, the current in each row being 4 amperes, or

132   ELECTRIC HEATING.

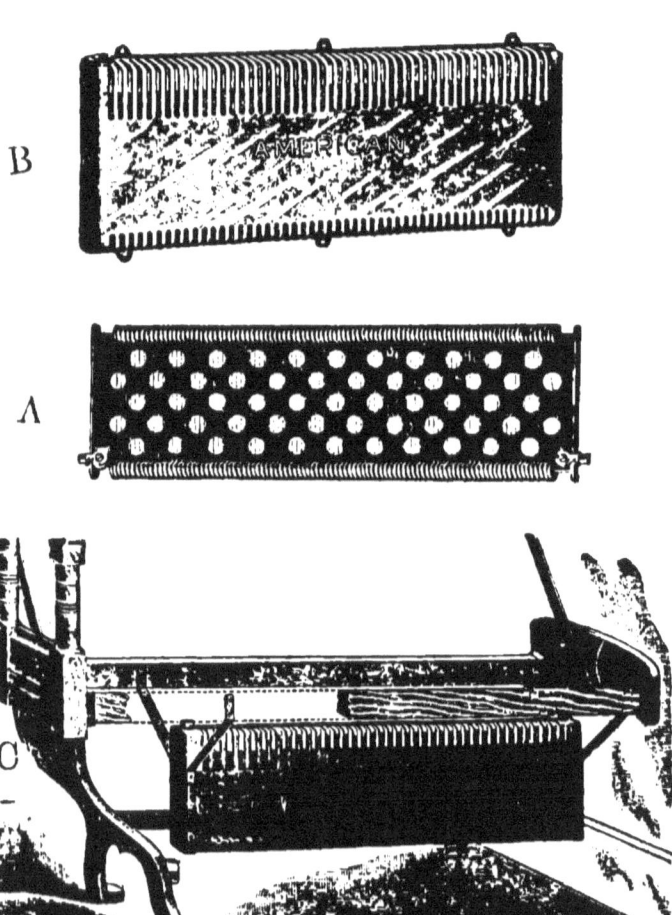

Fig. 29.—Car-Heater.

ELECTRIC HEATERS. 133

12 amperes in all, and the activity about 6 KW.

Another form of car-heater is shown in Fig. 29. Here the heating coil shown at A, consists of a wire wrapped in one long spiral around the insulated grid or frame. The heating coil is enclosed in a perforated iron cover shown at B, while at C,

Fig. 30.—Portable Air Heater.

the coil with its cover is shown in position below the car seat. Here the air enters the heater from the lower apertures and issues from those above, after passing over the heated wires.

Portable electric heaters, as their name indicates, are so constructed that they may be readily carried and temporarily attached in any room where electric supply is obtainable. These are made in a

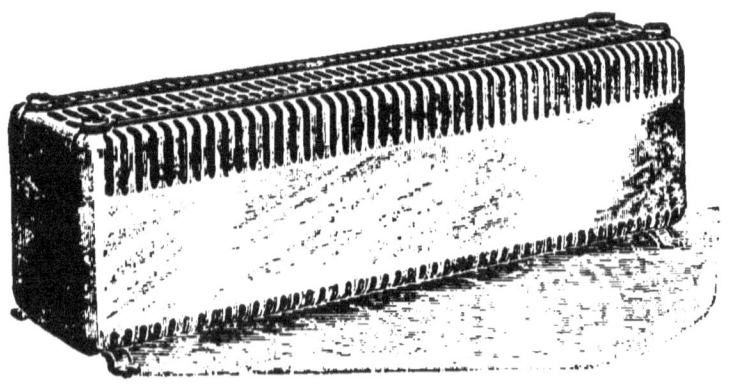

FIG. 31.—PORTABLE ELECTRIC HEATER.

variety of forms, but the principle in all cases is the same. A wire of suitable length and size is enclosed in the heater and free access given to it from the surrounding air. A form of cylindrical heater is represented in Fig. 30. Other

ELECTRIC HEATERS. 135

forms of portable heaters are shown in Figs. 31, 32, 33 and 34. That shown in Fig. 33 is 26 in. long, 7 in. in height, and 10½ in. wide, and is provided with three switches to regulate the temperature. A

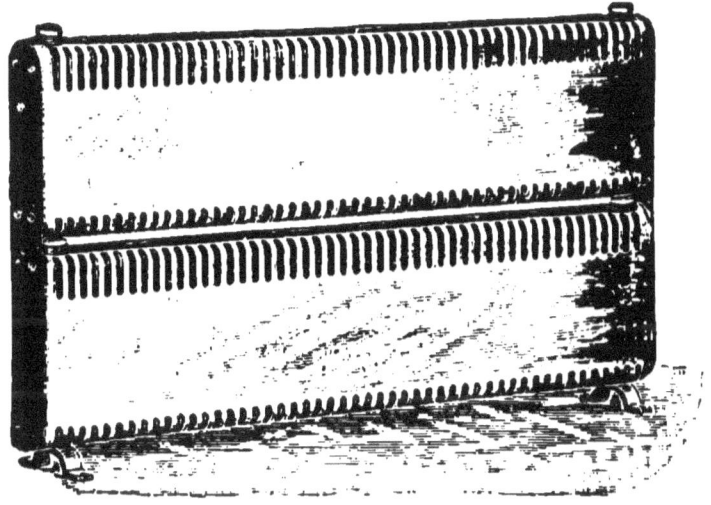

FIG. 32.—PORTABLE HEATER.

flexible attachment of the conductors to the heater is shown in Fig. 34. Fig. 35 represents a small stationary heater intended for attachment to a wall, corresponding, it may be, in position, to the ordinary hot-air register.

Figs. 36 and 37 show a form of electric heater suitable for office or house work.

Fig. 33.—Portable Heater.

Fig. 36 shows the exterior, and Fig. 37, the interior of the apparatus. The heating coils, six in number, are essentially of the same type as those employed in connection with the car-heaters represented

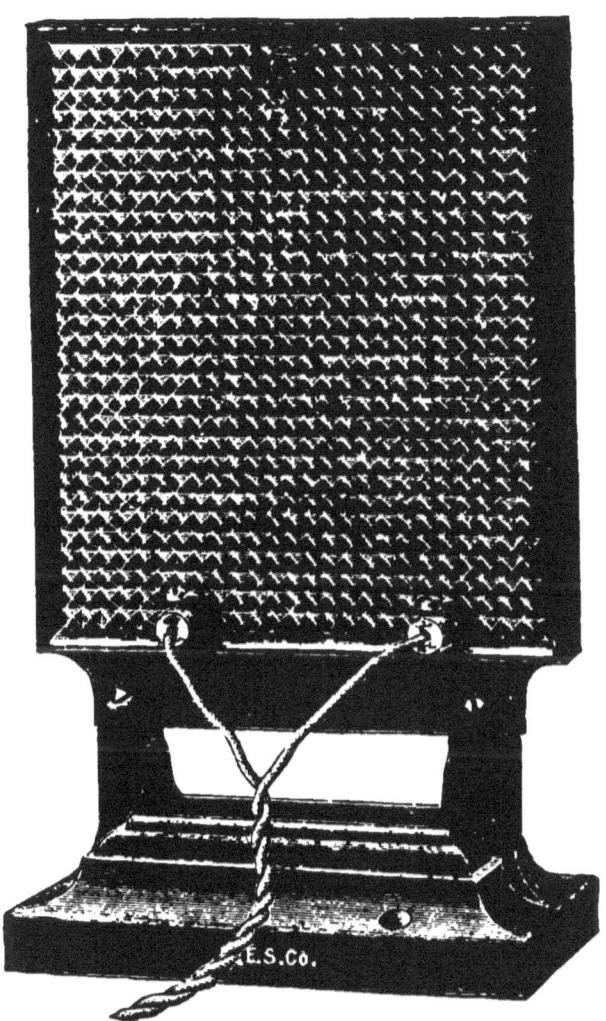

Fig. 34.—Atmospheric Heater.

in Figs. 23 and 24. The coils are wound on vertical porcelain frames, as shown in Fig. 37, and are sometimes provided with a temperature-regulating switch in such a manner that they may be connected in series, or parallel-series, and so produce less or greater activity. The stove case shown in Fig. 36, is made of Russia iron. The air enters at the bottom of the heater,

FIG. 35.—WALL HEATER.

passes up over the heated wire, and escapes at the top.

Electric air heaters may be employed for a variety of purposes, as, for example, for drying out the interiors of large caissons or tanks. A form of heater suitable for this purpose is represented in Fig. 38. It consists, as shown, of a number of coils,

ELECTRIC HEATERS.

FIG. 36.—PORTABLE ELECTRIC HEATER.

capable of being connected either in series or in parallel. It is 33 in. long, 12 in. wide,

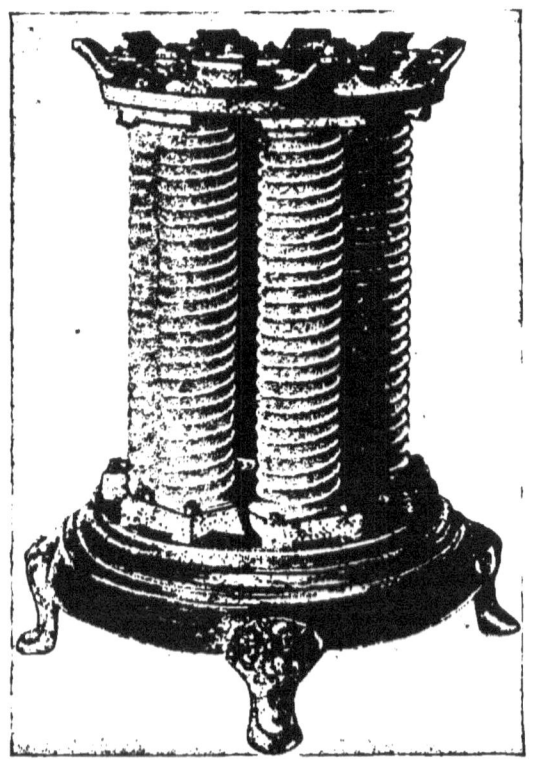

FIG. 37.—PORTABLE ELECTRIC HEATER, INSIDE VIEW,

7 in. in height, and is intended for a pressure of 110 volts with a maximum current

strength of 42 amperes; *i. e.*, a maximum activity of 4.62 KW.

As we have already seen, the product of the drop of pressure in a conductor and the current strength, equals the thermal activity in the conductor. Since in a heating coil, the drop is entirely of this

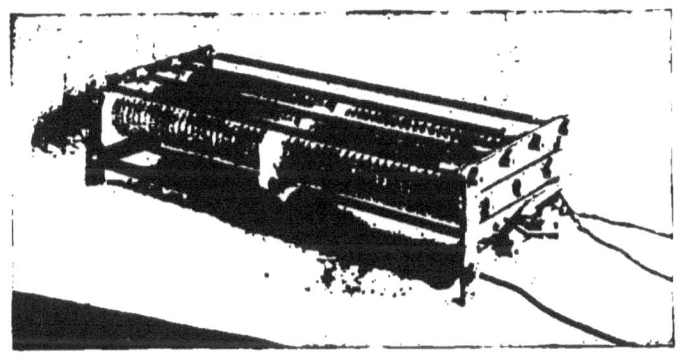

Fig. 38 Tank Heater.

nature, it is evident that all the energy of the current passing through the coil must appear in the circuit as heat, and all of this heat energy must be given to the external air on the cooling of the coil. Consequently, neglecting that small portion which is dissipated by conduction to the walls or floor, an electric air heater, as a device for converting electric energy into heat energy, may be regarded as a nearly perfect machine.

The cost of operating a car-heater will necessarily vary with the amount of activity developed in the car, and this, of course, will depend upon the number of amperes passing through the coils and the manner in which the coils are connected by the regulating switch. If, for example, there are four heaters in a car, and their resistance is 62.5 ohms each, then, when they are connected in series, the total resistance of the heating circuit will be say, 250 ohms. Assuming the pressure to be uniformly maintained at

500 volts, the current strength will be 2 amperes, and the thermal activity 1000 watts, or 1 KW. If the coils are connected in two rows of two each, the increased current which would flow through them would increase the resistance of each coil, by increasing its temperature, but assuming, for the sake of simplicity, that this increase of resistance is negligible, then the resistance of the coils, connected in two rows of two, will be $62\frac{1}{2}$ ohms, and a current of 8 amperes will pass, making the activity 4000 watts, or four times as great as in the preceding case. It is, of course, impossible to determine from these figures alone what the temperature in the car will be, since the air is being renewed by ventilation, and by the occasional opening of the car door. Moreover, the temperature produced will vary with the temperature of the external air, the

speed of the car, and with the direction and intensity of the wind. Consequently, in practice, it is necessary to provide for a variable production of heat so as to meet the requirements of a variable climate. It is found that the average amount of current required to warm the car, except in extremely cold climates, is three amperes at a pressure of 500 volts, or $1\frac{1}{2}$ kilowatts. The cost of a KW. hour, when supplied from a large power station to an extended system of cars, is usually a little over one cent and a half, per kilowatt-hour delivered. At this estimate, the average cost of heating a car in the winter is about 2.25 cents per hour, or 40.5 cents per car-day of 18 hours. The cost is stated to vary from 25 cents to 50 cents per car-day of 18 hours, according to the number of cars and the nature of the weather. It has been stated, from actual measurement in Boston, that cars having two doors, 12 windows and 850 cubic feet of space can be heated to an average temperature elevation of 25° F. above the external air during severe wintry weather by an expenditure of 2.5 KW.

Leaving out of consideration, however, the cost of the electric heating of a car, the advantages this method possesses over heating by a coal or oil stove are considerable. A stove fails to produce that uniform temperature so necessary to the comfort of the passengers, the centre of the car being more powerfully heated than the ends. The electric heater warms the air near the floor of the car, where warmth is most agreeable. Moreover, the electric heater requires practically no attention, does not necessitate the removal of dust, ashes or coal, and occupies no paying space. Consequently, where electric cars are used, the electric heater is coming into extended use, not only on account of its greater popularity, but also on account of its convenience.

When it is desired to apply heat directly to the surface of the body, for such medical treatment as would ordinarily employ hot water bags, the object can be much more conveniently obtained by a suitably constructed electric heater than by any method which depends for its heat on material warmed while away from its body, since, in all such cases, the cooling of the material necessitates its repeated renewal. An electric heater, suitable for local application to the body, and called a *flexible electric heater*, is shown in Fig. 39, because constructed of materials which enable it to be brought into intimate contact with the surface to

ELECTRIC HEATERS. 147

be heated. The heating coils B, are formed of German silver wire arranged as shown in the figure, placed on asbestos cloth and suitably insulated. The space

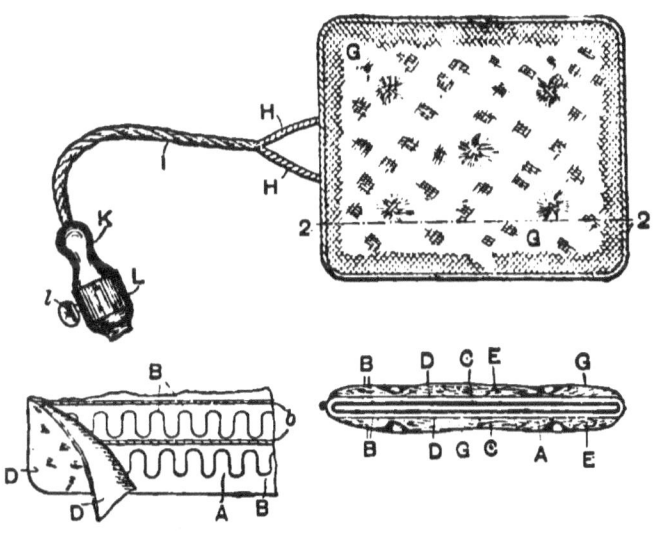

FIG. 39.—FLEXIBLE ELECTRIC HEATER.

surrounding the wires is filled with a solution of silicate of soda, which, on hardening, acts as a cement to hold the different parts together. A cushion, or flexible mass, is then made by packing mineral wool, or asbestos fibre, around the heating conductor and covering the mass with a suitable cover of cloth. The advantage of such a heater is that the heat can be readily maintained. The apparatus shown in the figure, ordinarily requires to be supplied with an activity of about fifty watts.

The electric heater has recently been adopted for the warming of the Vaudeville Theatre in London, England. The advantages of electric heating are specially marked in the case of theatres, where pure, warm air, without powerful currents or draughts are the desiderata. The heaters are two feet long and one foot wide. Twelve of these are attached to the skirtings round the walls, and twelve to the partition in front of the orchestra. Four large portable heaters are also employed with flexible attachments for use either in the centre of the theatre or at the sides. Each fixed heater takes a current of nearly 3 amperes, at 100 volts pressure, or develops an activity of nearly 300 watts, while the large, portable heaters develop 1200 watts. When all are working, the total activity is 11,400 watts or 11.4 kilowatts. It is stated, however, that, ordinarily, only two of the large portable heaters require to be used, so that the actual activity employed is 9 KW. The temperature of the auditorium is stated to be raised 20° F. by these heaters after they have been working for a reasonable length of time. The price charged being 8 cents per kilowatt-hour the cost of heating is 72 cents per hour, and to warm the theatre for four hours, $2.88.

It is similarly proposed to warm the stage by electric heaters to prevent the inrush of cool air into the auditorium when the curtain is raised.

To secure these results, it is only necessary to heat the air of the stage to practically the same temperature as that of the auditorium.

The advantages possessed by electric heating, already pointed out, are so marked in the case of the theatre, that with the general introduction of electric lighting into such buildings, their electric heating, either independently of or in conjunction with other methods of heating, is a possibility of the near future.

CHAPTER VII.

ELECTRIC COOKING.

ALTHOUGH, so far as its general electrical construction is concerned, an electric stove differs in no respect from an electric air heater, yet, there is this essential difference in the operation of these two pieces of apparatus; namely, that while the electric heater is so arranged as readily to impart its heat to a large volume of air in its neighborhood, the electric stove is so arranged that it can only impart its heat to a small volume of air confined in its interior. Consequently, for a given amount of heat produced, the air surrounding an electric heater acquires a temperature much lower than that within the stove.

Suppose any heating coil be taken, as, for example, the coil shown in Fig. 40, already described in connection with a car-heater in Fig. 24. Let us suppose that this coil has a resistance of 40 ohms (hot). If a current of three amperes be sent through it, the drop in the coil will be 3 × 40 = 120 volts, and the electric ac-

FIG. 40.—HEATING COIL.

tivity in the coil 3 × 120 = 360 watts, or nearly half a horse-power. This amount of heat is capable of raising the temperature of 20 cubic feet of air 1° F. per second. If this heater were placed at work in a closed chamber, the temperature acquired by the contained air would depend upon the volume of air. A large volume of air would acquire a lower temperature than a small volume of air. But the temperature attained would not depend only upon the volume of air in the chamber, but also upon the ability of the chamber to retain its heat, that is, to allow no heat to escape by conduction, radiation, or by convection, or open passages such as doors, windows, etc. For example, if the walls of the chamber were of cast iron, the temperature attained by the air within the chamber would be much lower than if the walls were thickly lined with some non-conductor, such as asbestos or felt. If, therefore, we know the volume of air in a chamber and also the rate at which heat escapes from it through walls or apertures, we have all the data necessary for the determination of the resulting temperature of the contained air.

An electric oven consists essentially of a small chamber, the air in which is practically isolated, the walls being nearly air-tight and lined with some non-conducting material, so as to retain the heat.

Fig. 41 shows a form of electric oven provided with a wooden external case, lined on the inside with asbestos or felt, and covered on the inside with bright, tin plate, which being a good reflector, tends to prevent heat from being conducted through the walls. Two electric heating coils are shown within at A and B, respectively, one at the top and the other at the bottom of the oven. By means of the switch, shown at the right hand of the drawing, either or both can be operated. A thermometer is inserted through a small hole in the top of the oven, to show the temperature of the contained air.

Fig. 42 shows another form of electric

ELECTRIC COOKING. 155

oven with three separate compartments and provided with a switch for operating

FIG. 41.—ELECTRIC OVEN.

the same. The large compartment is about 13 inches wide.

Fig. 42.—Electric Oven.

Fig. 43. Electric Coffee Heater.

158   ELECTRIC HEATING.

Fig. 43 represents a form of electric heater, suitable for heating a large quantity of coffee such as might be required

FIG. 44. ELECTRIC COFFEE-POT.

for use in a restaurant. Here the heater coil is situated in the base of the appa-

water-tight jacket.

Fig. 45.— Electric Kettle.

Fig. 44 represents a form of coffee-pot intended to be heated electrically from a pressure of 50 or 100 volts, absorbing, approximately, an activity of 500 watts. The electric heater coil is contained in the base of the pot. A flexible cord connects it with the nearest lamp socket.

Fig. 45 represents a form of electrically heated, four-quart tea-kettle. This kettle requires an activity or about 700 watts or nearly one horse-power, in order to boil one quart of water in ten minutes.

If one gallon of water be put into an electric tea-kettle, at say, a temperature of 41° F. (5°C.) and be raised, without actually boiling, to the boiling point, or 100° C., it would be elevated 95° C.; there would be, consequently, 3786 cubic centimetres elevated 95° C., (one gallon containing 3786 cubic centimetres) or $3786 \times 95 = 359,575$ water-gramme-degrees-centigrade of heat produced. But one *calorie*, or a water-gramme-degree-centigrade, requires an expenditure of 4.18

joules, so that the work required to be done in raising a gallon of water to the temperature of its boiling point, would be 359,575 × 4.18 = 1,503,000 joules. The cost of electric power in large quantities is usually about 8 cents per kilowatt-hour (*i. e.*, one KW. supplied for one hour, or 3,600,000 joules), and, in very small quantities, 15 cents per kilowatt-hour.

At 8 cents per KW. hour, the cost of raising one gallon of water to the boiling point would be $3\frac{1}{2}$ cents. At 15 cents per KW. hour, the cost would be $6\frac{1}{4}$ cents. This assumes, however, that all the electrically developed heat is utilized in raising the temperature of the water, which of course, is not the case since some heat is lost. For example, if we start with cold water in a cold kettle, the metal in the kettle will have to be heated before its heat can be communicated to the water, and, although in an air heater, any heat, so absorbed in the mass of metal of the heater would be returned to the air; in a water heater, this would not necessarily be returned to the water heated; beside, during the time required for the heating of the water, which would be about fifteen minutes for one gallon, the air outside the kettle would be warmed and would carry away some of the heat. The proportion of useful heat developed to total heat developed; or, as it is called, the *efficiency* of the kettle, would probably be about 70 per cent. Therefore, the actual cost of heating a gallon of water would be, approximately, $3\frac{1}{3} \times \frac{100}{70} = 4\frac{3}{4}$ cents at 8 cents per kilowatt-hour, or nearly 9 cents at 15 cents per kilowatt-hour.

It is evident, from the preceding figures, that at the present price of electric power, the electric water heater could not be economically employed on a large scale. It is to be remembered, however, that these prices are for power obtained from a central station generating electricity from coal, through the intervention of steam engines, boilers and dynamos. With water power, the cost would, probably, be much less, and even with steam power, where it is employed under the particular conditions applying to street-car driving, on a large scale, the cost to the central station of a KW. hour is only about $1\frac{1}{2}$ cents.

The cost of power developed for street-car propulsion is less than that of power developed for electric lighting for several reasons. Among others, to its being more continuously used, and to its being manufactured on a larger scale for street railway purposes than for lighting purposes.

Fig. 46 represents a form of electric chafing dish in which the electric heat is generated from a resistance coil, placed in a water-tight compartment at the base, where the wires enter. The apparatus is designed to hold about one quart of water,

Fig. 46.— Electric Chafing Dish.

and requires to be supplied with an activity of about 500 watts.

Fig. 47 represents an electrically heated stewing-pan for holding two quarts and designed for a supply of 700 watts.

ELECTRIC COOKING.

It will be evident, from an inspection of the preceding figures, that, excepting the electric stove, all the different types of electric cooking apparatus are practically of the same construction. In each, an electric heating coil, embedded in a water-

FIG. 47.—ELECTRIC STEWPAN.

tight manner, in a suitable part of the apparatus, supplies the heat that would otherwise be obtained from the ordinary coal stove or range. For the sake, however, of showing the convenience with which an electric heating coil or coils can be made to serve the necessities of the culinary art, Figs. 48, 49 and 50, representing respectively an electric skil-

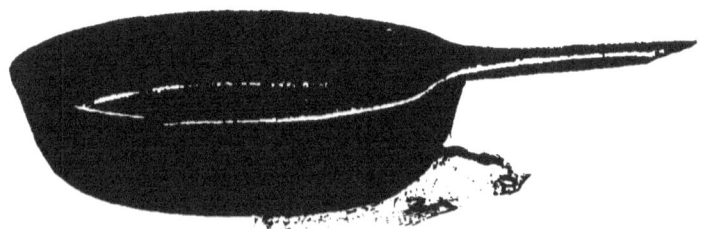

FIG. 48.—ELECTRIC SKILLET.

let, cake griddle and cooker, are shown.

In electric cooking apparatus contact with the supply mains is sometimes effect-

FIG. 49.—PANCAKE GRIDDLES.

ed by the ordinary screw plug. It is preferable, however, when much work of this character is to be done, to employ

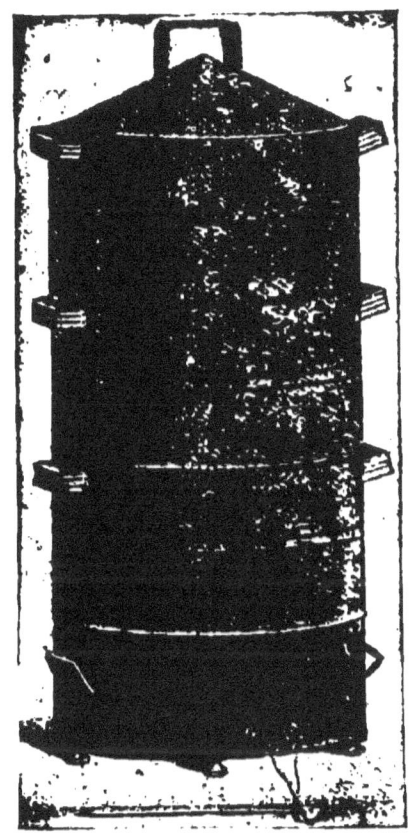

Fig. 50.—Electric Steam Cooker.

special connectors for this purpose. Two forms of *plug-switches* for such purposes are shown in Fig. 51. One of these is for attachment to the wall, and consists of a disc of wood, or hard rubber, with a slot containing a pair of separate springs connected with the supply mains. The insertion plug fits into the socket and connects two terminals from the flexible cord

Fig. 51.—Plug Switches.

leading to the heater with the spring clip, thereby establishing the circuit.

The other switch shows a very convenient method for connecting together two pairs of flexible cords. Each flexible cord terminates in a cylindrical block of wood or rubber in which is a pin and hole. The pin is connected with one terminal and the spring metal lining of the hole with the other terminal of the supply mains. The opposite plug is similarly fitted and the two are united by placing the pins into the respective holes and pressing the two together.

Although much remains to be accomplished in the way of improvements in electric cooking apparatus, especially in the direction of producing suitable heating coils that will last indefinitely without deterioration or short-circuiting, yet it will be evident that the advantages arising from the use of electricity in the kitchen are sufficiently great to warrant the belief that this practical use of electricity will rapidly grow. An ideal kitch-

170 ELECTRIC HEATING.

en, such as is capable of being furnished by apparatus already in existence, is

Fig. 52. —Electric Kitchen.

shown in Fig. 52. Here an electrically heated oven is provided with a hood,

ELECTRIC COOKING.

not to carry off the smoke of the fuel, but the odors from the cooking viands. A switchboard enables the utensils on the table to be connected with the supply mains as desired. B, is a hot-water boiler in which water can be readily heated electrically.

As we have already pointed out, the electric heater, considered as a device for transforming electric energy into heat energy, may be regarded as an extremely efficient apparatus. This cannot be asserted to the same degree of electric cooking apparatus, since, in such apparatus, some of the heat is lost; *i. e.*, diverted from the material to be cooked, and supplied to the surrounding metal, air or water. Since, however, all electric heat is usually obtained by burning coal in a central station, the cost of the electric heat on a large scale is considerably greater than the cost of the heat necessary for the same amount of cooking by the direct use of fuel in an ordinary range.

The larger the scale on which cooking is carried out, the greater the economical advantage of an ordinary fuel range over an electric range.

Under all circumstances, however, the electric heater is the more convenient and the more cleanly apparatus, and, when employed on a small scale for cooking, is often more economical than a coal range. Consider, for example, the case of preparing a cup of coffee by electric heating. Here, there is only required the generation of an amount of heat slightly in excess of that required to bring the water to the boiling point. Contrast this with the amount of fuel required to bring a cooking range to the temperature at which it can boil water. As regards convenience everything is in favor of the

FIG. 53.—SIMPLE ELECTRIC HEATER.

electric heater, since it requires only the closing of an electric circuit, which may be even done from another room, while bringing the range into use, requires the lighting of a fire.

A simple form of electric heater is represented in Fig. 53. Here the heat is obtained from an incandescent lamp, of size proportionate to the requirements of each case. As will be seen, the lamp is placed inside the hollow bottom of a coffee pot or kettle, which is blackened so as to absorb the heat. In this way 75 per cent. of the heat liberated by the lamp is utilized in the heating of the water. It is claimed that in the form shown, a 50-candle-power lamp, of say 200 watts activity, will heat $2\frac{1}{2}$ pounds of water to the temperature of boiling point in 25 minutes, and that when the water is at its boiling point it can be maintained at this temperature by the activity of a 16-candle-power lamp (about 50 watts), and in some cases even less.

Beside the uses we have already

ELECTRIC COOKING. 175

pointed out, of comparatively small electric currents for heating in connection

FIG. 54.—ELECTRICALLY HEATED GLUE-POT.

with heaters in cooking apparatus, a number of others might be mentioned. For example, Fig. 54 represents an electrically heated glue-pot, with a switch at the base, whereby the strength of current may be regulated within certain limits. This apparatus requires 700 watts for a one quart size, and 500 watts for pint

Fig. 55. —Electric Sad Iron.

size, when heated at the maximum rate. A much smaller activity is necessary to keep the glue hot when once melted.

Fig. 55 represents a sad iron, requiring about 250 watts for its operation, Fig. 56, a sealing-wax heater, and Fig. 57, a curling-tong heater. The sad iron is operated by a flexible cord attachment, but some

FIG. 56.—SEALING WAX HEATER.

forms are made in which the sad iron is free from electric connections and is merely laid upon an electrically heated

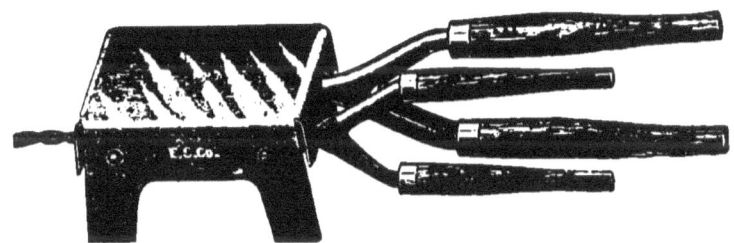

FIG. 57.—ELECTRIC CURLING-TONGS HEATER.

plate in order to acquire its heat by conduction.

As an illustration of what can be effected in the direction of electric cooking we may mention a banquet recently given in London, England, by the directors of an electric lighting company, to 120 guests, in which all the cooking was performed electrically. They were ten courses, which required for their preparation a total expenditure of energy of 60 kilowatt-hours, or on an average of one half a kilowatt-hour per guest.

The above company has notified the public that they will charge 8 cents per kilowatt-hour for cooking. Consequently, this would place the expense of such a banquet at 4 cents per guest for the ten courses. Considering the convenience of the process this charge cannot be regarded as exorbitant.

An electrically cooked banquet was not a possibility in the time of Franklin, yet

a banquet at which electricity played no insignificant part is thus humorously described by him in a letter written in 1769:
"Chagrined a little that we have been hitherto able to produce nothing in the way of use to mankind; and the hot weather coming on, when electrical experiments are not so agreeable, it is proposed to put an end to them for this season, somewhat humorously, in a party of pleasure on the banks of the Schuylkill. Spirits, at the same time, are to be fired by a spark sent from side to side through the river, without any other conductor than the water; an experiment which we some time since performed, to the amazement of many. A turkey is to be killed for our dinner by the electrical shock, and roasted by the electrical jack, before a fire kindled by the electrical bottle; when the healths of all famous electricians, in En-

gland, Holland, France, and Germany, are to be drank in electrified bumpers, under the discharge of guns from the electrical battery."

It may be of interest to our readers to note in this connection, that Dr. Franklin was not devoid of imagination, as may be gathered from a remark he makes concerning the turkey and other birds so killed:

"He conceited himself that the birds killed in this manner ate uncommonly tender."

CHAPTER VIII.

ELECTRIC WELDING.

In the proportioning of electric coils designed for heaters and cooking apparatus, care is taken that the electric resistance is such that, with the electromotive force employed, the resulting current strength should not be such that the coils shall reach an unduly high temperature. In no form of such apparatus are the coils allowed to reach an incandescent temperature; *i. e.*, a temperature at which they glow, or begin to emit light. There are, however, some very notable applications of the heating power of an electric current in which very high temperatures are employed, which we will now discuss.

These are capable of being divided into two sharply marked classes; namely,

(1) Those in which a metal forming part of an electric circuit is raised to its welding temperature; that is, a temperature considerably below the melting point of the metal.

(2) Those in which metals, or refractory substances, form portions of an electric circuit, and a temperature is obtained as high as is possible to produce under the circumstances, this temperature at times being the high temperature of the voltaic arc.

Apparatus of the first type find their examples in various forms of welding apparatus; those of the second type, in electric furnaces.

By the welding of two pieces of metal is meant causing them to strongly cohere, or hold together as a single piece, when powerfully pressed together. Some few metals, like lead, for example, possess the power of welding when cold. Thus, if two freshly-cut surfaces of lead, free from grease or oxide, are firmly pressed together, they will cohere so strongly that the welded joint may be as strong as other portions of the metal. Other metals, such, for example, as iron, copper, gold and steel, cannot be caused to cohere or weld in the cold by any pressure that can readily be brought to bear on them. If, however, these metals be heated to their welding temperature, generally a temperature at which they become incandescent, and then pressed together, either by quiet pressure, or by the blow of a hammer, they readily weld and cohere.

In order that welding may take place it

is necessary that the surfaces of the metallic weld be clean and free from oxides or other impurities. Such clean surfaces are insured by the use of a suitable flux, as, for example, borax, which removes the film of oxide that so readily forms on the surfaces of glowing metal.

In the practical welding of one metal to another, it has been found that the most efficient welding is obtained when a certain temperature is reached but not exceeded. In welding, carried on by means of the heat of an ordinary fire, the operator generally judges as to when this temperature is reached, by the color or appearance the metal acquires, and much of the welder's art consists in his ability to recognize precisely when the proper temperature has been reached in order to ensure the most effective joint.

The process of electric welding does not differ in any mechanical point from the welding of metals by the ordinary heating process, save, only, that the heat applied to the welding joint is of electrical origin, and, instead of the welding surfaces being separately heated in a furnace, and subsequently brought together, with the opportunity that their exposure to the air affords for the formation of a film of oxide over the surfaces to be united, in the electric process the surfaces are first heated by the passage of an electric current through them while placed in contact; and, when the welding temperature has been acquired, which even for large masses of metals requires only a few moments, are then suitably pressed together and the weld is affected.

The electric process of welding is not

only more convenient and rapid than the ordinary process, but by its means, metals have been effectively welded, which it is impossible to weld by the old process. By the application of the electric welding process not only can the ordinary metals, such as iron, steel and copper, be readily welded, but many metals which required under the old process to be previously bronzed, or covered by a layer of brass or solder, can now be directly welded. The following metals, for example, have been successfully welded electrically; viz., wrought iron, copper, gold, lead, zinc, tin, silver, aluminum and cast iron, and some of these metals have even been welded one to another.

The practical efficiency of any welded joint, of course, lies in the extent to which the tensile strength of the welded cross-section equals that of the unwelded portions of the bar. Judged by this test, an electrically welded joint possesses a marked advantage over an ordinary welded joint. Tests on the tensile strength of welded bars have shown generally that the bar is as strong at the welded joint as at other cross-sections, which is far from being the case in bars welded by the ordinary process, since the difficulty in applying the heat uniformly, and welding the bar promptly, without the formation of a deleterious scale, is greater in the case of an ordinary weld.

The current employed in electric welding may be either continuous or alternating. The amount of heat liberated in a given resistance, by a given current strength, is the same whether the current be continuous or alternating, although large bars, especially of iron, offer a greater resistance to the alternating than to the continuous current.

It is possible, therefore, to employ alternating currents for electric heating and this is, indeed, a very fortunate circumstance, since, when dynamo-electric machines are employed as the electric source, the use of the commutator is thereby obviated; for alternating-current generators employ no commutator, while continuous-current machines necessarily employ one. The enormous current strength employed in welding large bars, sometimes as high as 50,000 amperes, would necessitate the use of massive and expensive commutators, while with the use of alternating currents these are dispensed with.

Extended practical experience in the

welding of metals, especially in large masses, has demonstrated the fact that not only does no inconvenience attend the use of alternating currents in welding, but that, on the contrary, such currents actually possess advantages over continuous currents. In order to obtain a good joint, a certain temperature must be attained by the welding surfaces and this temperature should be as nearly uniform as possible. With the use of the continuous currents employed in such cases, the loss of heat at the surfaces of the metal causes the central portions of the mass to attain a higher temperature, thus rendering it more difficult to obtain a good welding joint, over a large cross-section. By the use of alternating currents, however, a more uniform distribution of temperature over the cross-section of the welded surfaces is obtained; for, although as before, the bar necessarily loses its heat from the surface, yet, as is well known, alternating currents tend to develop a greater heat at the surface of a large mass than at the central portions, and there is thus ensured a more uniform heating of the contact surfaces. Consequently, most welding processes are now carried out by the use of alternating currents.

The apparatus employed in electric welding may be divided into two classes; namely, those in which the alternating currents employed are generated directly from a specially designed alternating-current dynamo, and second, those in which the currents employed are taken from the secondary coil of a *step-down transformer*, that is, a transformer in which the secondary terminals supply a large current at a lower pressure, or a transformer in which the primary consists of a long, thin

FIG. 58. —DIRECT WELDER FOR WELDING BABY CARRIAGE TIRES AT THE RATE OF 1500 TO 2000 IN TEN HOURS.

wire, and the secondary of a short, stout wire. The first method is called the process of *direct welding*, and the second, that of *indirect welding*.

Fig. 58 shows a direct *welder* employed for welding the iron tires of baby carriages. Such a machine can make 1500 to 2000 welds in ten hours. It consists of an alternating-current dynamo, or alternator, with two field magnets M, M, and the armature A, revolving between the two pole-pieces, one on each side as shown at P^1. The armature is driven by a belt and pulley Y. The armature has two windings. One is connected with the commutator C, at the end of the shaft, and the brushes B, B, carry off a continuous, or commuted current, to the field magnet coils M, M, for their excitation. The other winding on the armature consists of a single massive turn of copper cable. Its extremities are brought to the collecting rings R, R, upon which rest heavy brushes to carry the powerful welding current to the two clamps $P, P,$ mounted above the platform $F F F$. These clamps can be caused to approach or recede by the turning of the handle h. The clamps $P, P,$ hold the two rods $d, d,$ which are to be welded together. The alternating currents, generated in the single turn of cable on the armature, are carried directly to the rods which are brought into end-to-end contact by the movement of the clamps. Since the clamps are attached to the rods to be welded close to the welded ends, it is evident that only the portions between these clamps and the welded surfaces, receive the welding current and attain the welding temperature. Moreover, in the immediate neighborhood of the clamps the heat is conducted away into the large metallic masses around the clamps. On the application of the current, the ends of the bars to be welded are pressed steadily together and the pressure is increased as the temperature rises. The current strength employed in the welding circuit seldom exceeds 4000 amperes, and the E. M. F. in the circuit is only two or three volts.

In order to avoid the use of collector rings for carrying off the heavy welding currents, forms of direct welders have been devised in which the armature is stationary, and the field is movable. In this case, the ends of the heavy cable, wound over the armature, are carried directly to the welding clamps.

Another form of direct welder is illustrated in Fig. 59. This welder is specially adapted to the purpose of welding strip-iron into hoops. Some of these hoops are represented at the bottom of the figure

Fig. 59.—Direct-Welding Apparatus.

with their welds at W. One of the magnets of the alternator is shown at M. The pole-piece P, embraces the armature, which is driven by a belt on the pulley Y. The whole machine can be moved forward with the aid of the rachet handle H, so as to tighten the belt, when necessary. The rheostat R, enables the strength of the current from the commutator C, to the magnet coils M, to be readily controlled. On the platform $F F$, are mounted the clamps $p\,p$, connected with the ends of the turn of cable on the armature, through the collector rings r, r, and brushes resting on the same. The strip S, to be welded, rests on the supports T, is then cut off at the right length and the two ends forced under the clamps p, p.

Whenever large bars or rods are to be welded, indirect welders are used. Any alternating-current machine can be employed for this purpose. The machines usually employed give an E. M. F. of 300 volts, with a frequency of 50 cycles, or 100 alternations; *i. e.*, 100 reversals of current per second. This alternating E. M. F. is led to the primary coil of a step-down alternating-current transformer, and the secondary coil of this transformer is brought directly to the bars to be welded. The E. M. F. in the secondary circuit varies from 1 to 4 volts, according to the character of the work to be performed, the strength of current required, the melting point of the metal to be welded, the size of the clamps, etc.

The connections of such an indirect welder are represented in Fig. 60. The alternator A, is driven by a belt B. A small belt b, from the same shaft drives the exciter *E*, which supplies the current for the field magnets of the alternator *A*, through the controlling rheostat *R*. The current from the collecting rings of the alternator armature, is carried through the

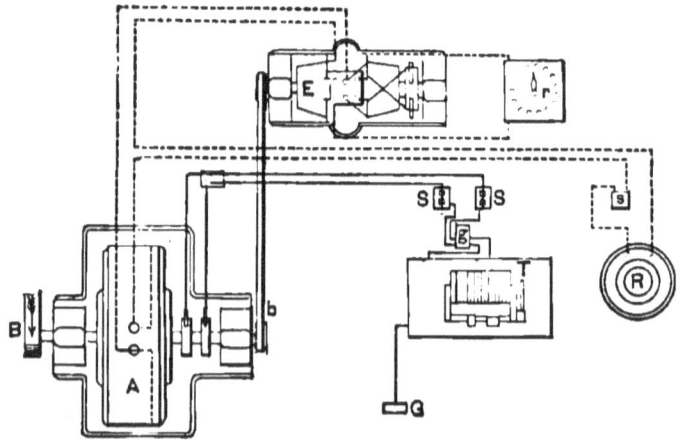

FIG. 60, —CONNECTIONS OF INDIRECT WELDING.

switches *S*, *S*, and the register *g*, to the primary coil of the welding transformer *T*. A register is employed to count the number of welds that the machine makes. The metallic mass or shell of the transformer T, is grounded by the ground connection G, in order to prevent any shock

Fig. 61.—Constant Potential Dynamo.

from being accidentally obtained from the apparatus, by the operator.

Fig. 61 represents an alternator, with six poles, intended for indirect welding. Here a separate continuous current generator G, supplies current to the magnets M, M, of the alternator. For this reason the alternator is called a *separately-excited* machine. There is, however, on the alternator shaft, a commutator c, which serves to commute part of the current from the armature A, and supplies this rectified or commuted current to the field magnets, in order to compensate for the drop in the pressure at the terminals of the armature, when the machine is running at full load. The machine is, therefore, said to be *compound-wound; i. e.*, contains two separate windings in its field magnets. The rings r, r, carry the current to the primary coil of the welding transformer at a pressure of about 300 volts. The handle H, is for tightening the belt on the main pulley P, by driving the generator forward on the guides g, g. This machine has a capacity of about 60 KW., or, at 300 volts pressure, will give a current of about 200 amperes.

Without entering into a minute explanation of the function of an alternating-current transformer, it is sufficient to state that it consists essentially of two coils of wire, placed side by side, called respectively the *primary* and *secondary coils* surrounding a core of laminated iron. One of these coils consists of many turns and the other of a few turns. In the case of the *welding transformer*, the secondary coil consists of a single turn of very heavy copper. When a rapidly *alternating current*, that is, a current which is rapidly changing its direction, is sent through the primary coil, currents, alternating equally rapidly, are generated by induction in the secondary coil. The relation existing between the E. M. F. that is caused to act on the primary coil and the E. M. F. produced by induction in the secondary coil, will depend upon the relative number of turns or loops of wire in each. If, for example, the primary contains 100 turns and the secondary a single turn, then, if the E. M. F. impressed upon the primary coil from the machine above described be 300 volts, there will be induced in the secondary coil an E. M. F. of about three volts. But, since the resistance of this single turn of very heavy copper is exceedingly low, the resistance of the secondary coil may be, say $\frac{1}{7000}$th of an ohm. The current strength which would flow through the secondary circuit might, therefore, be 21,000 amperes, a current that would necessarily possess large heating power; namely, $3 \times 21000 = 63,000$ watts activity.

Fig. 62 shows a form of welding transformer. A core *I*, of laminated iron, made up of a number of thin sheets piled together, is looped with a massive copper casting $S\,S\,S\,S$, which serves as a single

Fig. 62.— Welding Transformer.

turn of secondary conductor slit between the clamps C,C, as shown. Within the groove formed by this secondary casting is placed an insulated coil of wire forming the primary coil, but not shown in the figure. This transformer is, in reality, double, a second transformer being placed at the back, and only part of which is seen. Its construction, however, is identical with that just described. When an alternating electric current is sent through the primary coils, powerful currents are set up by induction in the heavy single copper turn forming the secondaries of these transformers as soon as their circuit is closed through the clamps and bars to be welded.

For very large work, which it would be impracticable to bring to the transformer, the transformer is so designed that it can be readily brought to the work. For this purpose the form of transformer shown in Fig. 63 has been devised. The outer shell

S S S S, of this transformer, is a copper casting made in two halves bolted together, serving as the secondary coil, and containing within it the primary coil. By this means

Fig. 63.— Large Welding Transformer.

the primary coil is protected from injury. Insulation is maintained not only by insulating the wire of the primary coil in the usual way, but also by filling the interior of the copper box with oil. The iron core, not shown in the figure, is linked both with the primary and secondary coils through the opening O.

In order to decrease the skill required for making an effective welded joint, the *automatic welder*, Fig. 64, has been devised. Here the proper degree of pressure between the contact surfaces is automatically applied, amounting for copper to 600 lbs. per square inch of welding cross-section, 1200 lbs. per square inch for iron, and 1800 lbs. per square inch for steel. The rods to be welded are placed in the clamps C, C_1, and are pressed together by the action of the weight W. The transformer T, supplies from its secondary coil the current strength required for effecting the weld. The movement of the clamps C, C_1, as the weld is effected, causes a contact to be made under the control of the screw *K*, actuating the

FIG. 64.—AUTOMATIC WELDER.

magnet *M*, which interrupts the main current.

Indirect welders are made in a variety of forms. Generally, however, the apparatus is protected from dirt, dust and injury by a suitable casing. A form of *automatic welder* is shown in Fig. 65, which is intended for the welding of copper wire.

The amount of power, which must be expended in effecting a weld, depends both upon the material and upon its cross-sectional area. If we double the cross-sectional area, we increase the amount of work to be expended by about 150 per cent. that is to say, we more than double the necessary expenditure of work. In order to weld bars of iron and steel one square inch in cross-section, nearly one *megajoule;* i. e., nearly 1,000,000 joules must be expended, or somewhat more than $\frac{1}{4}$ of a KW. hour. For a weld in brass, of one square inch in cross-section, about the same amount of work is

ELECTRIC WELDING.

Fig. 65.—Automatic Welder.

required; *i. e.*, a trifle more than one megajoule, and for a weld in copper one square inch in cross-section, an expenditure of nearly one and one-half megajoules is required.

Fig. 66 shows a form of welder intended

FIG. 66. – WELDER FOR CARRIAGE TIRES.

for welding carriage tires. The welding transformer is situated in the interior of the box upon which the clamps are mounted. Here the pressure is applied hydraulically from the cylinder G, under the action of the handle H. The tire to be welded is gripped in the clamps C, C.

Fig. 67.—Universal Welder.

Fig. 67 shows a universal welder adapted to a variety of work, and of 40 kilowatts capacity, so that at 2 volts E. M. F., the full-load current would be approximately

20,000 amperes, or 20 kilo-amperes. In this apparatus, as in the preceding, the pressure is applied hydraulically from the piston *G*, under the control of the handle

FIG. 68.—WELDER FOR CARRIAGE AXLES.

H. The handles *h*, *h*, are for operating the clamps *C*, *C*.

Fig. 68 shows a form of welder suited for welding wagon and carriage axles.

Fig. 69 shows a welder for steel wire cable or for bars of iron or steel.

Fig. 70 represents a form of welder

for welding steel spokes to their hubs. A circular platform is mounted above the transformer, as shown, and the four clutches grip as many spokes at a time. Water is supplied, through the

FIG. 69.—WELDER FOR CABLE OR BARS.

flexible pipes shown, to the upper clamps, which are hollow, so as to keep their temperature from becoming excessive under constant use.

Fig. 71 shows in detail some welds ef-

Fig. 70.—Welder for Wheel-Spokes.

ELECTRIC WELDING.   215

fected by the preceding apparatus. Here the advantage possessed by an electric weld for telegraphic joints becomes apparent. According to the old process as

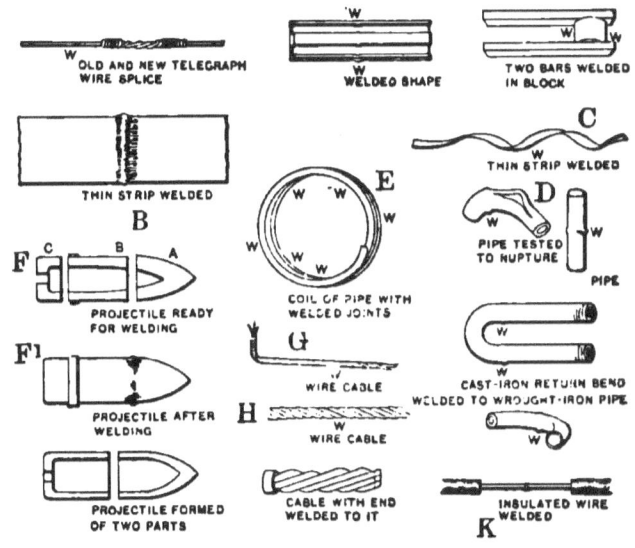

FIG. 71.—SPECIMENS OF ELECTRIC WELDING.

shown on the left in the illustration, the wire was twisted and soldered as indicated while according to the new method of welding, the ends of the wires are abutted and welded together, as shown in the lower right hand portion of the cut.

The extent of telegraphic welds may be inferred by the fact that a single firm, manufacturing telegraphic wire, makes on the average 600 welds daily by this method. At B and C, are shown thin strips welded together. At D, is shown a welded pipe which has been tested to the bursting point and which has burst not at the weld, but beyond it. At E, is a coil of pipe containing welded joints; at F, a projectile made in segments and ready for welding; at F^1, the same projectile after welding; G and H, wire cables welded; at K, an insulated wire with a welded joint. Of course, such welded joints can only be effected conveniently in the factory, as the welding apparatus is not usually available in the field.

Fig. 72.—Welder for Shrapnel Shells.

Fig. 72 shows a special form of welding machine for welding the hard steel points of shrapnel shells to their soft steel bodies. This is an operation that would be very difficult to accomplish by any

Fig. 73.—Pipe-Bending Apparatus.

other method. Fig. 73 represents a form of welding apparatus designed to heat a short length of pipe to enable the same to be readily bent. The pipe is held in the screw clamps C', C, and the current is sent through the short length of pipe between them, which is thus raised to the welding temperature except in the immediate neighborhood of the clamps.

In the system of street passenger railways, where the cars are driven by electric motors which take their current from trolley wires and tracks, a necessity exists for ensuring a continuous electric contact between the separate rails constituting the tracks. This is effected, in practice, by connecting the abutting ends of the rails by means of stout copper wires, or *bonds*, as they are termed. No little difficulty has arisen in practice, owing to the imperfect contact thus ensured between the surfaces of the bond wires and the rail, a considerable resistance being introduced into the circuit of the rails from this lack of good connection, as well as from the liability to corrosion through galvanic action. Not only is a continuous conductor necessary for the economical transmission of electric current over the line, but also to reduce to a minimum the electric corrosion of the gas and water-pipes, or other masses of metal situated along the line in the neighborhood of the railroad tracks. Again, unless the contact between adjoining rails is electrically good, the advantages gained by buried cables, or *ground feeders*, to constitute a return circuit, is materially diminished.

An attempt has been made to overcome these difficulties by rendering the entire length of rail constituting the track one continuous metallic conductor. This is done by welding the abutting ends of the rails together, while in place, on the track. Since the rails are welded in place, the electric current has, of course, to be carried to the weld. To this end, the necessary welding appliances are placed on a special car which either takes its current from the trolley wire, or from any alternating-current circuit that may be in the neighborhood. When the continuous current from the trolley wire is employed for this purpose, the pressure being approximately 500 volts, this current drives a *motor-dynamo*, or *rotary transformer*, placed on the car, and by this means the continuous current received from the trolley is converted into an alternating current and afterward delivered into the primary coil of the welding transformer.

The car employed for this purpose is shown in Fig. 74. The welding transformer, with its large clamps, is seen suspended from a beam at the rear end of the car. The same transformer is shown more clearly in Fig. 75, which represents the welding transformer in place, in actual work upon a track weld. By means of a

Fig. 74.—Track-Welding Car.

motor in the car, the surface of the rails is ground by a revolving grinder for a few inches on each side of the joint, so as to prepare a clean surface of iron on which the weld is to be produced. Two iron

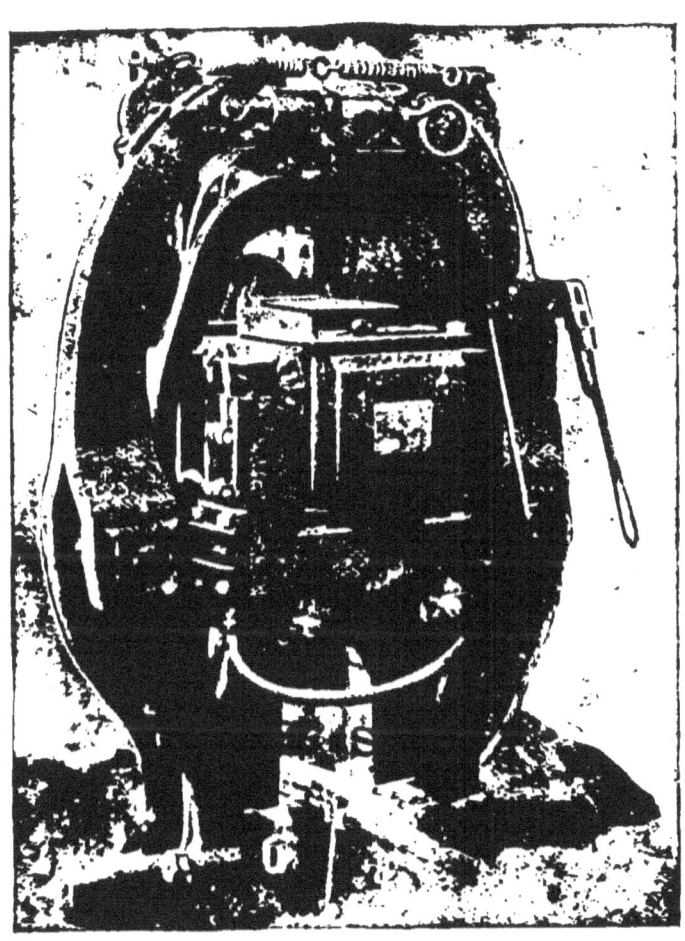

Fig. 75.—Track Welder at Work.

chucks are then placed in position, one on each side of the joint, and the electric current is forced from the jaws of the welder through the chucks and across the two ends of the rails. By this means the chucks and rail ends are brought together up to the welding temperature. Hydraulic pressure is exerted upon the chucks by the hand pump P, shown on the right. When the weld is effected, the two chucks and the two ends of the rail form one solid mass. The massive secondary copper casting, or single turn $S\ S\ S\ S$, is represented in the figure with its two lower extremities $S_1\ S_2$ forming the terminals which are brought into contact with the chucks. The primary coil is contained within the secondary shell or box, and the laminated iron core $I\ I$, is passed through or linked with both. The two heavy iron jaws $J\ J,\ J\ J,$ pivoted at V, are

ELECTRIC WELDING. 225

drawn apart by the spiral springs at the top, but are forced together by the hydraulic pump M, so as to bring pressure transversely upon the chucks through the heads of the secondary terminals $S_1 S_1$. It will be seen, therefore, that the rails are not pushed together, end to end, but are welded transversely.

Fig. 76 represents the appearance of a welded rail, after the operation is completed. The area of this weld is from 12 to 16 square inches. The current strength required from the trolley wire may reach 275 amperes, representing an activity of about 137.5 KW. This is delivered from the motor-dynamo, or rotary transformer, as an alternating current at a pressure somewhat in excess of 300 volts, and, after allowing for the losses of power in the rotary transformer,

226   ELECTRIC HEATING.

as well as in the welding transformer, about 120 KW. can be delivered to the

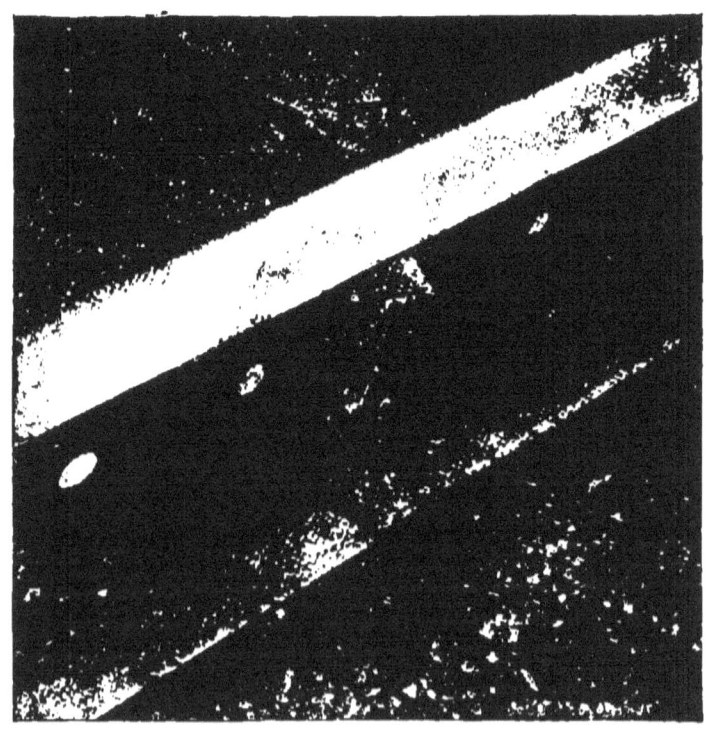

Fig. 76.—Welded Rails.

track, representing a current strength of very nearly 60,000 amperes. The welding transformer is oil-insulated, so that the whole apparatus can be worked in the rain. Water is circulated through the jaws, in order to cool them when at work. Under favorable circumstances, four joints can be made per hour.

A street rail, weighing 70 lbs. per yard, when prevented from expanding and contracting, owing to the entire rail being in a single length, requires about 150,000 lbs. tensile strength to withstand the stresses produced in it by the expansions and contractions, following changes in temperature due to the seasons. An electric weld requires more than 250,000 pounds to break it. Consequently, a track is not likely to break at a weld owing to the stresses produced by temperature variation. It is necessary, however, in practice, to keep the track firmly from bending in summer, by securely fastening it to the sleepers.

In order to cite an example of the practical application of electric track welding, it may be mentioned that in the city of Boston, four miles of Providence girder street car rails, weighing 61 lbs. per yard, were electrically welded in the summer of 1893 in one continuous length. It had been the general belief, up to the date of this experiment, that a track so welded could not resist the tendency of its own expansion and contraction to pull it to pieces. These four miles remained in good condition until the following winter, when they broke in about 80 places, but, in nearly all cases, it is interesting to note that these fractures did not occur at the joints, but about four to eight inches from them. These fractures were repaired by being electrically welded. The track lasted intact through the summer of 1894, but again broke the following winter in about 30 places. It is a curious fact that these breaks did not occur at regular intervals, but several would usually appear within a few feet, and then none, perhaps, for half a mile. It is claimed that the difficulty referred to in the preceding paragraph can now be overcome.

In all the methods of welding thus far described, a single process is employed; namely, the parts to be welded are brought into contact and a powerful electric current is sent through the contact surfaces until they are raised to the welding temperature. The temperature is never allowed to reach the fusing point. Another method of welding, which differs radically from the preceding, consists, practically, in bringing the metals to be welded to the fusing point. This is accomplished by the use of the voltaic arc as follows; one terminal of the source of current, preferably a storage battery of between 50 and 100 volts E. M. F., is connected to the metals to be welded, and the other terminal, to a rod of hard carbon, which is brought into contact at the welding surfaces and then separated a short distance from them, so as to form an arc between the metal and the end of the carbon electrode. By this means, a partial fusion is obtained, which results in an electric *soldering*; or, as it is sometimes called, a welding at the joint. This method of uniting the ends of metal bars or rods, is not unlike the *burning process* as applied to lead, in which two abutting surfaces or ends of lead sheets are united by the aid of a blow-pipe flame. It is evident that this method is not capable of as many applications as is the method previously described, since the heat, being only superficially applied, is incapable of giving to joints of any considerable cross-section, that uniformity of temperature on which a good weld is dependent. The process, however, possesses some advantages, and has been successfully applied to the filling of blow holes in castings. It is evident that masses of metal introduced at the fusing temperature into such blow holes, under the action of the electric arc, tend to render the mass of metal fairly homogeneous, provided the precaution has been taken to previously heat the casting to a dull redness.

The same process has been applied to *longitudinal welding*, or *calking* of plates that have already been riveted, in order to make a water-tight joint and instead of employing a calking tool. As before, however, the process is limited to the case of comparatively thin plates.

Another method, also dependent on the heat of the voltaic arc, consists in deflecting, by the aid of an electromagnet, the arc existing between two carbon points and directing the flame against the surfaces to be welded. This apparatus constitutes, in fact, an electric blow-pipe.

CHAPTER IX.

ELECTRIC FURNACES.

THE intense heat of the voltaic arc, forming, as it does, the most powerful source of heat known, led many investigators, at a very early date, to apply it in various metallurgical processes. These processes were, as a rule, carried out in what may be properly styled *electric furnaces*. That is, in furnaces, the heat of which was obtained electrically, either by means of the voltaic arc, or by the heat of intense incandescence of such refractory substances as graphite or carbon. It may be well to point out, in this connection, that the electric furnace differs radically from any furnace in which the heat is obtained by ordinary combustion, in that means must necessarily be provided, in the combustion furnaces, for carrying off the products of combustion. This not only ensures an inefficient form of furnace, but also necessitates the cooling or chilling of the furnace by the loss of heat, and by the ingress of cold air. In marked contrast with this, in an electric furnace, no essential gaseous products of combustion are formed in the production of the heat, and, consequently, all the heat developed is retained, with the exception of such losses as occur through the walls of the furnace by conduction. Electric furnaces have been known in the art as early as 1848, and since that time have been very frequently employed.

The electric furnace assumes a variety of forms, one of which is shown in Fig. 77.

ELECTRIC FURNACES. 235

Here a voltaic carbon arc is employed as the source of heat, the arc being permitted to play in the interior of a crucible of refractory material, surrounded by a non-conducting mass, usually of fire-

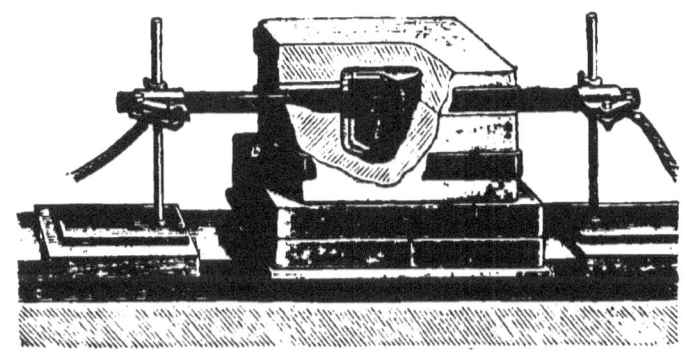

FIG. 77. —ELECTRIC FURNACE.

brick. Since comparatively little heat escapes by conduction, the temperature which may be attained in the interior is exceedingly high. This particular form of furnace was employed to ascertain the temperature at which carbon boils.

Although constructed in a variety of forms, all electric furnaces may be divided into two classes; namely, first, those in which the operations carried on are effected by means of the intense heat electrically produced, and, second, those in which the operations are effected by *electrolysis; i. e.*, the power possessed by an electric current, under certain conditions, of effecting chemical decompositions. By far, however, the greater number of commercial electric furnaces belong to the first class.

In all electric furnaces the heat is obtained either by means of the electric arc or by electric incandescence. Since carbon is one of the most refractory substances known, it is generally employed either as the material between which the arc is formed, or as the substance for leading the current into the furnace. Since, as is well known, the carbon arc is the most intense source of artificial heat we possess, and the peculiar construction of the electric furnace permits this heat to be readily accumulated, the temperature reached is the highest artificially obtainable. Consequently, under these conditions, chemical processes become possible on a commercial scale, that heretofore could only be conducted on a small scale in laboratory research.

As an example of a commercial process carried on under the intense heat of the electric furnace, we may mention the manufacture of a compound of silicon and carbon, known in commerce as *carborundum*. This material is carbon silicide, a molecule of which consists of an atom of silicon united to an atom of carbon. This product is of considerable commercial value in the arts, owing to its great hardness, and is extensively used as an abrasive material, as a substitute for emery and corundum, and has even been employed

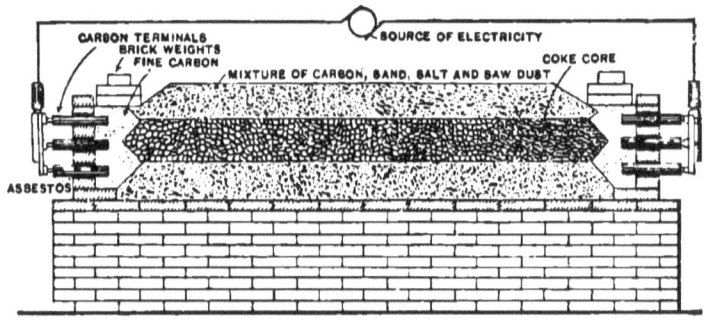

Fig. 78.—Longitudinal Section of Carborundum Furnace.

in the place of diamond dust, for the polishing of gems.

The furnace employed for the production of carborundum is shown in longitudinal section, as charged ready for the passage of the current, in Fig. 78. It consists substantially of a rectangular chamber, whose walls are formed of brick and fire-clay. The furnace chamber is charged with a central core of granular coke, surrounded by a mixture of carbon, sand, salt and sawdust. In order to effectively connect the electric source with the central carbon core of the charged furnace, carbon rods or terminals are placed at each end of the furnace and brought into good electrical connection with the core by means of a filling of fine carbon tightly packed around them. When a powerful electric current is sent through this furnace, a chemical action occurs, under the influence of the intense heat, whereby a combination is effected between the carbon mainly of the central core and the silicon of the sand, with the formation of a silicide of carbon called carborundum.

A cross-section of the furnace, prior to the passage of the current, is shown in Fig. 79, and another cross-section, after the passage of the current, in Fig. 80. Reference to the latter figure will show

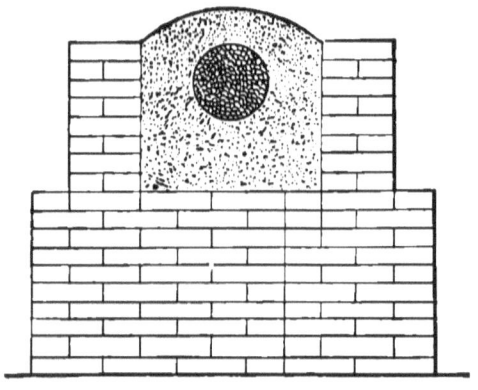

Fig. 79.—Section Through Furnace before Passage of Current.

that a portion of the coke core still remains unaltered, while carborundum in the crystalized and uncrystalized states surrounds this unaltered core.

Another commercial application of the

ELECTRIC FURNACES. 241

electric furnace in which the product is obtained by high temperature, is in the process for the manufacture of calcium carbide. In this process the product is obtained by the prolonged action of an

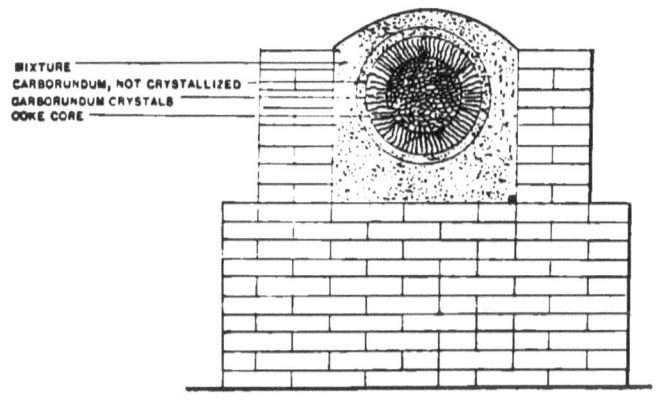

FIG. 80.—SECTION THROUGH FURNACE AFTER PASSAGE OF CURRENT.

electric arc on a mixture of lime and carbon, placed inside a suitably formed smelting furnace, formed of refractory materials. The form of the furnace is shown in Fig. 81. The outer shell A, consists of a cylindrical fire-brick cover or bench, inside of which is placed a crucible B, of carbon or graphite. Both the cruci-

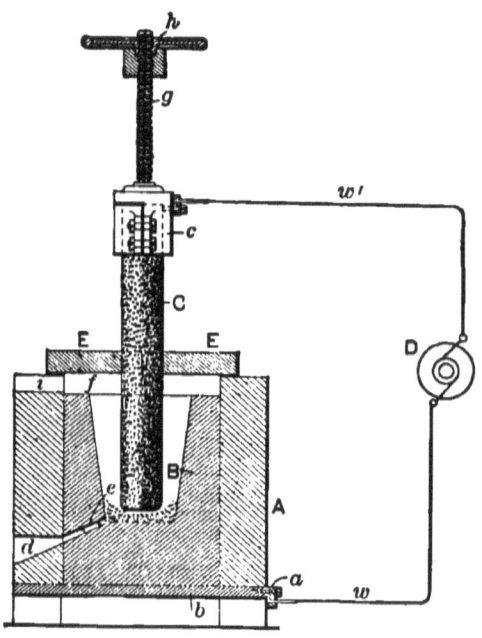

FIG. 81.—FURNACE FOR PRODUCTION OF CALCIUM CARBIDE.

ble B, and the masonry A, rest on a conducting plate b, of metal, to which one of the terminals of the dynamo is connected, the other terminal being connected to the carbon bar or pencil *C*, forming the movable electrode of the furnace. The furnace is provided with the cover *E*, formed of a single or double carbon plate. This is insulated from the body of the furnace *B*, by means of a plate of non-conducting material *F*. The material to be acted on is placed at the bottom of the furnace, and heat applied by means of a current passing between the electrode *C*, and the crucible *B*. A screw-thread shaft *G*, attached to the carbon, permits the adjustment of the central electrode in the nut *h*. A tap hole is provided at *d*, for discharging the products of the furnace from time to time. During operation, this hole is closed by a plug of clay or other suitable material.

An alternating current of from 4000 to 5000 amperes under a pressure of from 35 to 25 volts, representing an activity of about 135 KW., or 180 H.P., can, it is claimed, produce daily in such *furnaces* a yield of one short ton, or 2000 pounds of calcium carbide at a cost of about $15.

No little attention has recently been attracted to the preparation of calcium carbide, from the fact that when thrown into water, it is capable of yielding acetylene gas, a combination of hydrogen and carbon ($C_2 H_2$), which possesses a high illuminating power when burnt in air. Either a continuous or an alternating current may be employed in its production. One of the most important uses to which acetylene can be applied is the enrichment of ordinary illuminating gas, so as to increase its light-giving power.

Up to the present time, perhaps, the most important application of the *electric*

ELECTRIC FURNACES. 245

furnace is to the production of aluminium, either pure or alloyed with copper.

Fig. 82 represents a section of an electric furnace which produces aluminium bronze alloy; *i. e.*, aluminium alloyed with copper. This furnace consists essentially of a rectangular chamber of fire-brick

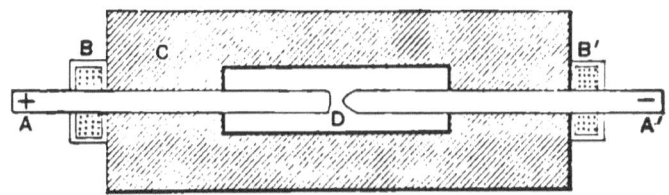

FIG. 82.—ELECTRIC FURNACE FOR THE PRODUCTION OF ALUMINIUM ALLOYS.

provided with carbon electrodes entering the charged chamber.

A convenient size for such a furnace has an interior length of five feet, a width of one foot, and a height of one foot. The charge occupies the centre of this space in a mass roughly 3 feet long, 7 inches wide and 3 inches high, the space between the charge and the wall being filled with limed charcoal. The furnace employs a carbon arc as the source of heat, the arc being formed between the carbon electrodes which lead the current through the furnace. In the figure these electrodes are shown at $A+A-$, the arc being formed between them at D. The electrodes pass through openings in the ends through boxes B, B^1, filled with granulated copper. The charge in such a furnace is frequently a mixture of 50 lbs. granulated copper, with 25 lbs. of crushed cryolite, a mineral rich in aluminium, and 12 lbs. of charcoal. The current strength varies from 1200 to 1500 amperes, and is maintained at a pressure of about 50 volts for 5 hours. Under these circumstances, the ore of aluminium is reduced in the presence of highly heated carbon, and the reduced metal enters into an alloy with the molten copper. When the furnace is cleared, 50 lbs. of alloy are obtained having from 15 to 35 per cent. of aluminium and a small quantity of silicon.

In another process, by means of which the aluminium is obtained in a pure state, the decomposition is effected by electrolysis. Here the current is led through an electrolytic bath of alumina dissolved in a double fluoride of aluminium and potassium, maintained in a fused state by the heat evolved during the passage of the current. In one process in which this is effected, the crucible, which consists of an iron box suitably lined with carbon forming the cathode or negative electrode, is charged with the ores of aluminium, and a carbon rod, standing vertically in the centre, forms the anode, or positive electrode. The current enters by this carbon rod, and, after passing through the materials of the furnace, leaves it at the negative or external surface by means of the iron frame suitably connected to the other pole of the dynamo. The current strength employed is about 3500 amperes at a pressure of approximately 35 volts, representing an activity of 122.5 KW. The furnace is so arranged that the metal can be tapped off and withdrawn as it is formed, so that the process is a continuous one, fresh ore being added from time to time. The effect of the current is not only to keep the charge in the furnace molten by the heat produced in the passage through the furnace, but also to reduce the metal from the ore by electrolytic action. By these means the metal obtained is very nearly pure. The iron box is usually cubical in shape, and is two feet deep. It has an opening beneath, which is supplied with a plug of carbon or clay to permit of the pouring off of the metal.

The electric furnace has been employed in obtaining a number of rare metallic substances among which chromium may be mentioned.

In the use of electric furnaces for metallurgic purposes many advantages arise from the fact that a vacuum can readily be maintained within the furnace during the operation. For this reason metals obtained in the fused state from their ores by electric reduction, or metals fused in air-tight furnaces by the application of heat of electric origin, produce sharper and much more homogeneous castings than those melted when exposed to the air. Moreover, such castings are devoid of troublesome blow holes and blasts, and are denser than ordinary castings.

In one form of electric furnace, the ore is not only reduced to the metallic state, by the action of the current, but is also cast directly from the furnace, within which a vacuum is maintained. This furnace consists of an air-tight chamber, provided with an inclined hearth, arranged so as to permit the reduced and molten metal to flow directly from the furnace into the mould when so desired. The chamber of the furnace is filled with a suitable mixture of ore, flux and reducing agent, and subjected to the influence of the electric current; or, the furnace is given a charge of the metal to be melted and a current applied sufficient to melt it, while in the presence of a vacuum.

The practical limit of size proposed for such a chamber is 40 feet in length and capable of holding $1\frac{1}{2}$ tons of metal at a charge. By working such a chamber with a current of about 30,000 amperes, at fifty volts pressure; *i. e.*, at an activity of about 1500 KW., somewhat less than the activity already employed in the aluminium electric furnace at Neuhausen, the entire charge can be fused and run off in about a quarter of an hour. Such a furnace would, therefore, be capable of turning out a very large number of castings in a single day.

It might be supposed that the electric melting of metals would be more expensive than the ordinary method employing the regenerative furnace, but, bearing in mind the fact that all the heat developed by the electric current can be liberated exactly where it is wanted, and that the loss of heat in such a furnace is very small, it is evident, that even where water-power is not obtainable, this method might compete with coal on a commercial basis. For example, it has been estimated that in order to smelt a short ton of iron in the Siemens-Martin regenerative furnace, from 1000 to 1400 pounds of coal are required. By the electric process here described, assuming that coal is burned to drive the dynamo and operate the air pump employed in maintaining the vacuum, the same work can, it is claimed, be done by the consumption of from 720 to 800 lbs. of coal.

In the use of a furnace of the above type for the direct production of pig iron from iron ore, the resulting iron can be made to contain much less carbon that in that produced by the ordinary blast furnace, since the ingredients can be much more closely proportioned in the electric furnace than in the ordinary blast furnace. Experiments made have produced pig iron containing less than 3 per cent. of total carbon.

The electric furnace has been employed for the artificial production of very small diamonds. When carbon is melted and vaporized in the electric furnace, it condenses in the form of graphite with the specific gravity of about 2. Indeed this same process occurs in every arc lamp, the carbon being volatilized at the positive electrode, a portion of this vapor condensing in the form of a nipple of graphite on the cooler, negative or opposite electrode. In order to produce the diamond, great pressure is necessary. This can be obtained by forming a solution of carbon in molten iron, and allowing the iron to solidify suddenly, thereby bringing sufficient pressure upon the contained carbon to crystalize the latter into diamonds. A molten solution of carbon and iron, obtained in an electric furnace, is suddenly poured into lead that has just been separately melted. The iron and carbon, being lighter than molten lead, float to its surface in the form of globules, and solidify. These globules, when dissolved in suitable acids, will leave as a residue the diamond crystals which are unfortunately very minute, but have all the physical properties of larger natural gems.

CHAPTER X.

MISCELLANEOUS APPLICATIONS OF ELECTRIC HEATING.

BESIDE the different commercial applications of heat of electric origin, which we have already described, there are others of great interest that would appear to have a reasonable probability of coming into extensive use in the near future. We will, therefore, devote the consideration of the closing chapter to some of the more interesting of these applications.

In the manufacture of *harveyized armor plates*, now extensively employed on warships, considerable difficulty has arisen in drilling the plates so as to permit them to be riveted together. The harveyized steel plate, as is well known, is so extremely hard, that the ordinary drill has no effect whatever on it. Attempts have been made to soften, or anneal, these plates at the points where the drill holes have to be made, but although the intense heat of the oxy-hydrogen blow-pipe has been tried for this purpose, it has been found to be insufficient. For this reason a strip around the edges of the plate had to be left unhardened, so as to permit of the drilling, and this was an element of weakness. It has been found, however, that under the intense heat of the voltaic arc, even the harveyized plate was annealed, or restored to the soft condition, then readily permitting penetration by the drill. This method of *electric annealing* is carried out specifically as follows: Blocks of copper are laid on the surface of the plate and connected with an alternating current transformer, resembling a welding transformer. By this means, on the passage of the current, intense heat is developed in the plate between the two electrodes or masses of copper. The temperature is then slowly lowered by reducing the current strength. This has the effect of withdrawing the temper, or annealing the plate between the two blocks of copper. It has been found that alternating currents are more favorable for the concentration of the heating effect than continuous currents, a fact due to the inductance in the iron.

The heat of the voltaic arc has been employed in a process of electric casting already described and mentioned as a process of electric soldering. This process is applicable to the cases of repairing flywheels, steam cylinders, connecting-rods, etc. It consists essentially in the employment of the voltaic arc taken between two metal electrodes. One of the electrodes, consisting of the mass of the metal to be repaired, is fixed, and the other, the movable electrode, is made of the metal which is to be fused and employed in the repairing. Under these conditions, the arc is formed between the metal to be repaired and the metal employed in the casting or filling of the intervening space, the latter melting, and dropping into the interstices of the metal to be filled with the metal and then soldered or welded.

This process requires about 8 amperes per active square millimetre of the metal electrode. The usual diameter employed for the electric soldering tool is from 6 to 10 millimetres. It is necessary that the metal which receives the molten application should itself be raised to a red heat, as, otherwise, the molten metal introduced would chill too rapidly, and thus prevent an effective junction.

Probably one of the most valuable miscellaneous applications of electric heating is to be found in the various processes which have been designed for the *electrical working* and *forging of metals*. In these processes the metal is brought by heat of electric origin to the temperature required for its working, shaping or forging.

In this, as in other commercial applications of electric heating, one of the most marked advantages obtained is found in the fact that the heat is developed in the exact locality where it is needed, and not elsewhere; is developed only to the extent it is needed, and not to an unnecessary extent; and, moreover, only at the time when it is needed. Instead of requiring a long previous heating in the forge or furnace with a waste of fuel, the metal is quickly heated by the electric current. Moreover, heat of electric origin is capable of much finer and closer regulation than is heat of the ordinary forge or furnace. Then again, automatic devices may be readily introduced whereby the current can not only be controlled as to amount, but also can be cut off as soon as a certain temperature is reached. This will be found a matter of considerable advantage in cases where the metals to be worked require tempering, since the heat to which they are subjected can be made absolutely uniform, irrespective of the size of the piece to be heated. Moreover, the bar can be heated uniformly throughout all portions of its area of cross-section.

MISCELLANEOUS APPLICATIONS. 261

A decided advantage in electric forging lies in the rapidity with which the heating can be obtained; for, if the power applied be ample, the bar to be forged can be brought up to the forging temperature in less than a minute. At the same time it is to be remembered that no very large bars have yet been treated electrically. This process has so far been applied mainly to the production of comparatively small cross-sections of metal, although, of course, it is only a question of the amount of electric power to permit the process to be carried on in larger sizes.

The power required to heat an iron or steel bar one square inch in cross-section and 20 inches long is about 27 KW. and requires about $2\frac{1}{2}$ minutes, representing a total work done of about 4,000,000 joules or $1\frac{1}{8}$ KW.-hrs. = 200,000 joules-per-cubic-inch. A larger bar 3 feet long and 3 inches in diameter, would require about 75 KW. over ten minutes, or 45 megajoules =14 KW. hours, or nearly 180,000 joules-per-cubic-inch.

Two distinctly different methods are in use for obtaining the electrical heating of the material to be shaped or forged; namely, heating it by passing a sufficiently powerful current through it while in the air, and passing an electric current from it into a mass of surrounding conducting liquid. The former process, as in electric welding, requires the use of a powerful current strength at a low pressure and is best obtained by means of an alternating-current transformer. The latter process, on the contrary, requires comparatively small current strength, but a comparatively high electrical pressure.

MISCELLANEOUS APPLICATIONS.

FIG 83.—ELECTRIC FORGE AND ELECTRIC COOKING RANGE.

Fig. 83 represents the apparatus employed when the former method of heating is adopted. *T T*, is a large alternating-current transformer for reducing a current of comparatively high pressure to one of very low pressure, but of correspondingly increased strength. In the particular case represented the primary coil of the transformer receives about 40 KW. at full load at a pressure of 1500 volts and consequently a current strength of about 24 amperes. The secondary coil delivers nearly 40 KW. at full load at a pressure of about 4 volts and, consequently, with a current strength of about 10,000 amperes. The secondary terminals of the transformer are connected with the copper massive conductors 1 and 2; 3 and 4; 5 and 6; and 7 and 8; any pair being selected according to the character of the work to be heated. These conductors terminate beneath in clamps or holders suitable for different sizes of work. Bars to be heated are shown at *B*, bridging across the distance between the two electrodes or clamps. The attention of the reader is called to the electrical cooking range shown at the right, not because it has any connection with the forging process, but from the fact that it differs from the electric cooking ranges described in the earlier chapter of this book, since its heating coils are properly proportioned to produce the required temperature within it from a large current strength and a low pressure of, say four volts, instead of from a high pressure of perhaps 100 volts, and a correspondingly reduced current.

A number of samples of work done by the hammer on metal heated electrically

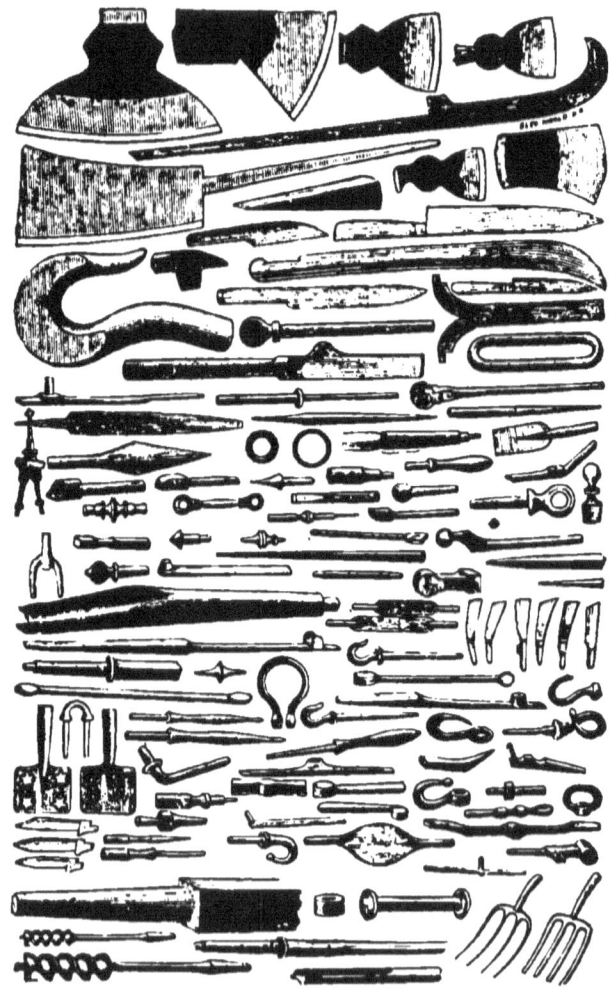

Fig. 84.—Samples of Forgings Electrically Heated.

by this process is shown in Fig. 84.

The second method for heating consists in plunging the metal to be heated beneath the surface of the conducting liquid, when held in a metal clamp connected with the negative pole of a continuous-current source of E. M. F. The metal to be heated is made the negative pole, and the vessel containing the liquid is provided with a metal lining of lead connected with the positive pole. Under these circumstances the current passes from the liquid to the metal to be heated. The current strength employed is sufficient to produce free electrolysis of the liquid with the production of free hydrogen gas at the surface of the metal to be heated, the high resistance of which causes so intense a heat at this surface as to practically set up an *electric arc* over its surface. The heat so produced rapidly penetrates the mass of the metal and raises its temperature.

It is to be observed that this method can only be employed with a continuous current. The heating process is conducted without any oxidation of the metal to be heated, its surface being thoroughly protected by the enveloping mass of hydrogen. The metal surface of the vessel containing the liquid becomes oxidized by electrolysis during the operation of the process, and has to be renewed from time to time. The main resistance in this liquid tank exists at the surface of the metal, in the film or layer of hydrogen, and, consequently, it is at this surface that the heat is almost entirely liberated. Consequently, the amount of current employed is automatically regulated by the surface area of the immersed metal, the larger the surface the greater the current strength which will flow. The pressure employed for such a liquid heater may be from 100 to 500 volts, and the current strength from 45 amperes upward.

In order to render the liquid conducting, a suitable conducting salt such as sal soda is dissolved in the water to a specific

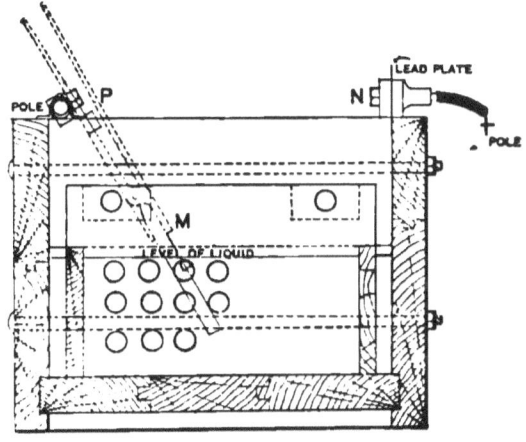

FIG. 85.—END VIEW OF HEATING TANK.

gravity of 1.2 at 84° F., and to every ten gallons of the solution five pounds of borax are added.

Fig. 85 represents an end view in cross-section of the tank employed. The pincers P, are connected with the positive pole of the source and hold the metal article M, so that this is partially submerged. The negative pole N, is connected with

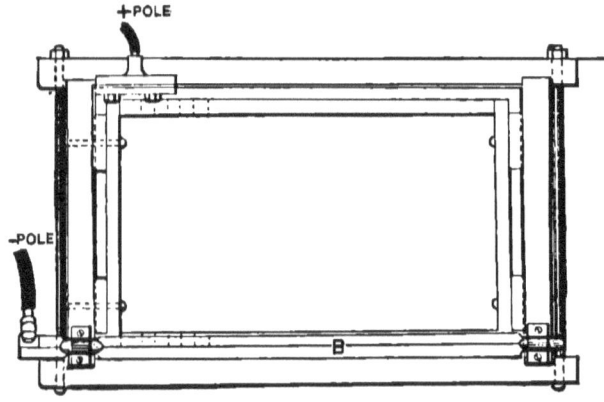

FIG. 86.—PLAN OF HEATING TANK.

the sheet lead lining of the tank. Fig. 86 represents the same apparatus in plan view.

THE END.

INDEX.

A.

Abnormal Temperature Elevation of Circuits, How Avoided, 87, 89.
Acetyline Gas, Illuminating Power of, 241.
———, Production of, from Calcium Carbide, 243, 244.
Activity, Definition of, 26.
———, Muscular, Obtained from the Sun, 15.
——— of Circuit, 41.
——— of Electric Circuit, 35.
——— of Laborer, 27.
———, Unit of, 26.
Aerial Bare Wires, Effect of Character of Surface on Temperature Elevation of, 66.
———, Effect of Extent and Surface on Temperature Elevation of, 66.
Affinity, Chemical, 9.
Air Heater, Portable Electric, 133, 134.
———, Resistance Offered by to Escape of Heat from Conductors, 59, 60.

Alloys, Effect of Temperature on Resistivity of, 56, 57.
——, Lead-Tin, for Fuse Wires, 91.
Alternating Current, Definition of, 201.
Alternating Currents, Advantages Possessed by, for Electric Heating, 189, 190.
Alternator for Indirect Welding, 199, 200.
——, Separately-Excited, 200.
Aluminum, Alloys, Furnace for Production of, 244-247.
——, Metallic, Electric Production of, 247-249.
Ampere or Coulomb-per-Second, 35.
Annealing, Electric, of Harveyized Armor Plates, 255-257.
——, Influence of, on Resistivity, 54.
Armor Plates, Harveyized, Electric Annealing of, 255, 256.
Atlantic Liner, Activity of Driving Engines of, 27.
Atmospheric Heater, 137.
Automatic Welder, 206-208.

B.

B. T. U., 29.
Back Electric Pressure, 41.
Banquet, Franklin's Electrically Cooked, 178, 180.
Bare Aerial Wires, Temperature Elevation of, 45, 46.
—— Conductors, Electrical Heating of, 37-68.

INDEX. 273

Block, Ceiling, 105.
———, Porcelain, 87–98.
———, Safety Fuse, 96.
———, Cut-Out, 106, 107.
Bond for Street Railways, 219.
Box, Cut-Out, 107.
Branch Fuse, 113, 114.
British Heat Unit, 29.
——— Thermal Unit, 29.
——— Thermal Unit, Value of, 29.
Buried Conductor, Permissible Temperature Elevation of, 85.

C.

C. E. M. F., 42.
———, Development of, by Motor, 45.
——— in Circuit, Distribution of, 42, 43.
Cable Welder, 213.
Calcium Carbide, Furnace for Manufacture of, 242, 243.
Calking, Electric, 231, 232.
Calorie, Lesser, 29.
Capacity, Carrying, of Conductor, 75.
Car for Direct Welding, 222.
——— Heater, Electric, 125, 126.
——— Heater Regulating Switch, 127, 129, 218.
——— Heating, Cost of, 142–145.

Carbon, Effect of Temperature on Resistivity of, 56, 57.
Carborundum, 237.
—— Furnace, 238–241.
Carriage Axle Welder, 212.
—— Tire Welder, 210, 211.
Carrying Capacity of Fuse Wires, 92.
Castings, Sharpness of, When Produced from Electrically Fused Metals, 251.
Ceiling Fixture, Fuse-Block, 104, 105.
Chafing Dish, Electric, 164.
Chemical Affinity, 9.
Circuit, Activity of, 40.
——, C. E. M. F. and Activity of, 46, 47.
——, Distribution of C. E. M. F. in, 42, 43.
——, Wires, Bare, 64.
——, Wires, Covered, 64.
Circular Mils, Definition of, 52.
Coal, Energy in Pound of, 10, 11.
——, Origin of Energy in, 12.
Coal-Beds, Store-houses of Solar Energy, 13, 14.
Coffee Heater, Electric, 157.
Coffee-Pot, Electric, 158.
Compound-Wound Machine, 200.
Conduction, Loss of Heat by, 63.
Conductor, Carrying Capacity of, 75.
——, Temperature Elevation of, 62.

INDEX.

Conductors, Pure Metallic Effect of Temperature on Resistivity of, 56, 57.
——, Transmission, Necessity for Maintaining Low Temperature of, 69.
Conduits for Insulated Wires, 77.
Connections for Indirect Welding, 198.
Convection, Approximate Amount of Heat Lost by Conductor per Foot of Length per Second, 67.
——, Loss of Heat by, 63.
——, Loss of Heat, Practical Independence of Extent and Character of Surface on Temperature Elevation of, 66, 67.
Convectional Losses in Conductors, Effect of Motion of Air on, 68.
Cooking, Electric, 151–180.
Copper-Tipped Fuse Wires, 95.
Cost of Car Heating, 142–145.
Coulomb, or Unit of Electric Quantity, 32.
Coulomb-Volts or Unit of Electric Work, 33.
Counter E. M. F., 41.
—— E. M. F., how Produced, 44–45.
Counter-Hydraulic Pressure, 40, 41.
Covered Conductors, Electrical Heating of, 69–86.
Curling-Tongs Heater, Electric, 177, 178.
Current, Electric, Work done by, 33.

Current Strength, Effect of, on Temperature Elevation of Wire 80.
—— Strength, Effective.
—— Strength, Thermal, 62.
Cut-Out Box, 107.
Cylindrical Electric Heater, 120–122.

D.

Diameters of Fuse Wires, 92.
Diamonds, Electric Furnace for the Production of Artificial, 253, 254.
Difference of Electrical Pressure, Electrical Flow Produced by, 32.
—— of Thermal Pressure, 72.
—— of Water Level, Liquid Flow Produced by, 31.
Direct Welder, 191–193.
—— Welder, Electric, 223.
—— Welding Apparatus, 194–196.
—— Welding Car, 222.
Dissipation of Heat, 9.
Doctrine of the Conservation of Energy, 20.
Drop, Definition of, 43.

E.

E. M. F., 35.
——, Counter, 41.
——, Impressed, 41.

INDEX. 277

Earth-Buried Conductors, Loss of Heat by, 81.
Economy of Electric Smelting, 252.
Effective Thermal Resistance of Earth-Buried Conductors, 81.
Efficiency of Electric Kettle, 152.
——— of Steam Engine, 11.
Electric Boiling of Water, Cost of, 161.
——— Car Heater, 125, 126.
——— Circuit, Activity of, 35.
——— Cooking, 151–180.
——— Cooking, Advantages of, 172, 173.
——— Heaters, 117–150.
——— Heater, Advantages Possessed by, for Car Heating, 124.
——— Heater, Advantages of, 119.
——— Radiator, 123.
——— Resistance, 38.
——— Source, Definition of, 24.
Electricity and Heat, Relations between, 19, 20.
——— Circumstances Regulating Flow of, 37.
Electrolysis, Definition of, 236.
Electrolytic Heating, 267–269.
Electromotive Force, Definition of, 32.
Elements of Work, 22.
Energy, Conservation of, 20.
——— in Pound of Coal, 10, 11.

Energy, of Coal, Origin of, 12.
—— Storage of, in Water Reservoir, 30.

F.

Falling Water, Storage of Solar Energy in, 16.
Fan, Electric, 154-156.
Feeders, Ground, for Electric Railways, 220.
Flexible Electric Heater, 146, 147.
Flow of Electricity, Circumstances Regulating, 37.
—— of Water, Circumstances Regulating, 37.
Foot-Pound-per-Second, Definition of, 26.
Foot-Pounds, 23.
Force, Definition of, 21.
——, Electromotive, Definition of, 32.
Forging, Electric, 263, 264.
——, Electric, of Metals, 259, 260.
——, Electric, Samples of, 266.
Franklin's Electrically Cooked Banquet, 178-180.
Full-Load Current, Temperature Elevation under, 80.
Furnace, Electric, 233-254.
——, Electric, Definition of, 233.
——, Electric, for Manufacture of Calcium Carbide, 242, 243.
——, Electric, for the Manufacture of Carborundum, 238-241.

INDEX.

Fuse-Box, Ceiling Fixture, 104, 105.
——— Boxes, Mica-Covered, 100, 101.
——— Boxes, Porcelain-Covered, 102, 103.
———, Branch, 113, 114.
——— Links, 94.
———, Main-Circuit, 113, 114.
——— Screw Block, 107.
——— Wire, Definition of, 89.
——— Wire Strips, 93.
——— Wires, 87-115.
——— Wires, Copper-Tipped, 95.
——— Wires, Carrying Capacity of, 92.
——— Wires, Composition of, 91.
——— Wires, Diameters of, 92.
Fuses, Safety, 90.

G.

Glue-Pot, Electrically Heated, 175.
Ground-Feeders for Electric Railways, 220.

H.

Harveyized Armor Plates, Electric Annealing of 255-257.
Heat and Electricity, Relations between, 19, 20.

Heat and Mechanical Work, Relations between, 17, 18.
―――― Conduction, 63.
――――, Dissipation of, 9.
――――, Loss of, by Conduction, 63.
――――, Loss of, by Convection, 63.
――――, Loss of, by Radiation, 64.
――――, Nature of, 8.
――――, or Molecular Oscillations, 9.
―――― Unit, British, 29.
――――, Unit of, 28, 29.
Heater, Cylindrical, Electric, 120-122.
――――, Electrical, Advantages of, 119.
――――, Electric Tank, 141, 142.
――――, Electric Wall, 138.
――――, Flexible Electric, 146, 147.
――――, Portable Electric, 140, 141.
Heaters, Electric, 117-150.
――――, Electric, Essential Construction of, 119.
――――, Electric, Requisites for, 119.
Heating, Electric Coil Conductor for, 122.
――――, Electric, Tank for, 269, 270.
――――, Electric, Miscellaneous Applications of, 255-270.
――――, Electrical, of Bare Conductors, 37-68.
――――, Electrical, of Covered Conductors, 69-86.

Heating of Conductor, Effect of Insulating Covering on, 69, 70.
——, Electrolytic, 267–269.
Hemp Covered Wires, Permissible Temperature Elevation of, 85.
Horse-Power and Kilowatt, Relative Values of, 36.
——, Definition of, 26 27.
Hydraulic Resistance, 38.

I.

Impressed E. M. F., 41.
Indirect Welder, 197, 198.
—— Welding, 192.
—— Welding, Connections for, 198.
Insulated Wires, Conduits for, 77.
—— Wires in Conduits, Temperature Elevation of, 77.
—— Wires, Mouldings for, 76, 77.
Insulating Covering, Effect of, on Electrical Heating of Conductor, 69.
—— Covering, Effect of Thickness of, on Temperature Elevation, 72.
—— Covering, Thermal Resistance of, 72.
International Unit of Activity, 26, 27.

J.

Joints, Welded, Tensile Strength of, 187.

Joule, 33.
———, Definition of, 24, 25.
——— per-Second, 26, 27.
———, Value of, in Foot-Pounds, 25.

K.

Kettle, Electric, 158.
———, Electric, Efficiency of, 162.
Kilowatt, 36.
Kitchen, Electric, 169-171.

L.

Laborer, Activity of, 27.
Law, Ohm's, 39.
Lead Sheathing of Wires, Influence of, on Temperature Elevation of, 74.
Lead-Tin Alloys for Fuse Wires, 91.
Lesser Calorie, 20.
Level, Electric, Difference of, 32.
Links, Fuse, 94.
Loss of Heat by Conduction, 63.
——— of Heat by Convection, 63.
——— of Heat by Radiation, 63, 64.

M.

Mechanical Work and Quantity of Heat, Relations between, 17, 18.

Megajoule, Definition of, 208.
Metallic Ores, Electric Production of, 250, 251.
Metals, Electric Forging of, 258-260.
———, Electrical Working of, 259-270.
Mica-Covered Fuse Boxes, 100, 101.
Microhm, Definition of, 48.
Mil, Definition of, 51, 52.
Mils, Circular, Definition of, 52.
Molecular Oscillations or Heat, 9.
Motor, Electric Development of Counter E. M. F. by, 45.
———, Dynamo, 221.
Mouldings for Insulated Wire, 76, 77.
———, Wooden, Rule for Size of Wire in, 77, 78.

N.

Nature of Heat, 8.
Negative Resistivity, Temperature Coefficient of, 56.

O.

Ocean Cables, Temperature Elevation of, 86.
Ohm's Law, 30.

P.

Pan-Cake Griddles, Electric, 166.
Physical State, Influence of, on Resistivity, 54.
Pipe-Bending Apparatus, Electric, 218.
Plug-Switch for Electric Heaters, 168, 169.

Porcelain-Covered Fuse-Boxes, 102, 103.
—— Fuse-Block, 97, 98.
Portable Electric Heater, 133, 134.
—— Electric Heater, 140, 141.
Positive Resistivity, Temperature Coefficient of, 56.
Pressure, Back Electric, 41.
——, Counter-Electric, 41.
——, Counter-Hydraulic, 40, 41.
——, Electric, Difference of, 32.
——, Hydraulic, 40, 41.
——, Unit of Electric, 33.
Primary Coils of Transformer, 201.
Purity, Influence of, on Resistivity, 54.

Q.

Quantity, Electrical, Unit of, 32.

R.

Radiation, Loss of Heat by, 63, 64.
Radiator, Electric, 122, 123.
Rails, Electrically Welded, 226.
Rate of Doing Work, or Activity, 26.
Reduction, Electric, of Metallic Ores, 250, 251.
Regulating Switch for Car Heater, 127–129.
Reservoir of Water, Activity in, 34, 35.

INDEX. 285

Resistance, Electric, 38.
——, Hydraulic, 38.
——, Thermal, of Insulating Covering, 71.
Resistivity, Definition of, 48.
——, Effect of, on Pure Metallic Conductors, 56, 57.
——, Effect of Temperature on, 56, 57.
——, Influence of Annealing on, 54.
——, Influence of Physical State on, 54.
——, Influence of Purity on, 54.
—— of Alloys, Effect of Temperature on, 56, 57.
—— of Carbon, Effect of Temperature on, 56, 57.
Rotary Transformer, 221.
Rubber Covered Wires, Permissible Temperature Elevation in, 85.
Rule for Size of Wire in Wooden Mouldings, 77, 78.

S.

Sad Iron, Electric, 176.
Safety Fuse-Block, 96.
—— Fuses, 90.
—— Strips, 93.
—— Transformer Fuse-Box, 109, 110.
Screw Block, 107.
Sealing-Wax Heater, Electric, 177.
Secondary Coils of Transformer, 201.
Separately-Excited Alternator, 200.

ELECTRIC HEATING.

Sharpness of Castings When Produced by Electrically-Fused Metals, 251.
Shrapnel Shells, Welder for, 217, 218.
Size of Wire in Wooden Moulding, Rule for, 77, 78.
Skillet, Electric, 166.
Smelting, Electric, Economy of, 256.
Socket Attachment, 107.
Solar Energy, Storage of, in Coal Beds, 13, 14.
———— Energy, Storage of, in Falling Water, 16.
———— Energy, Storage of, in Wind, 16.
Soldering, Electric, 230, 257–259.
Source, Electric, Definition of, 32.
Specimens of Electric Welding, 215.
Steam Cooker, Electric, 167.
Steam Engine, Efficiency of, 11.
Step-Down Transformer, 190.
Stew Pan, Electric, 165.
Street Railway, Bonds for, 219.
Strips, Fuse Wire, 93.
————, Safety, 93.
Subdivided Conductors, Temperature Elevation of, 75, 76.
Subway, Temperature Elevation of Wires in, 82.
Sun, Prime Source of Muscular Activity, 15.
Switch, Car Heater Regulating, 127.
———— for Electric Fan, 155.

T.

Table of Resistivities, 48, 49.
Tank for Electric Heating, 269, 270.
—— —Heater, Electric, 141, 142.
Temperature, Effect of, on Pure Metallic Conductors, 56, 57.
——, Effect of, on Resistivity, 56, 57.
——, Effect of, on Resistivity of Alloys, 56, 57.
—— Elevation of Circuits, Abnormal, How Avoided, 87, 89.
—— Elevation of Conductor, 62.
—— Elevation of Conductor, Effect of Thickness of Insulating Covering on, 72.
—— Elevation of Conductors in Conduits, 77.
—— Elevation of Ocean Cables, 86.
—— Elevation of Subdivided Conductors, 75, 76.
—— Elevation of Wire, Effective Current Strength of, 80.
—— Elevation of Wire, Maximum Time Required for, 83, 84.
—— Elevation of Wire, Safe, 79.
—— Elevation of Wires in Subway, 82.
—— Elevation Permissible in Hemp-Covered Wires, 85.
—— Elevation Permissible in Rubber-Covered Wires, 85.

Temperature, Elevation, Permissible, in Buried Conductors, 84, 85.
Tensile Strength of Electrically Welded Joints, 187.
Therm, Defination of, 29.
Thermal Current Strength, 62.
——— Resistance, Effective, of Earth Buried Conductors, 81.
——— Resistance of Insulating Covering, 71.
——— Unit, British, 29.
Tin-Lead Alloys for Fuse Wires, 91.
Transformer, Primary Coils of, 201.
———, Rotary, 221.
———, Safety Fuse Box, 109, 110.
———, Secondary Coils of, 201.
———, Step-Down, 190.
———, Welding, 201–206.

U.

Unit, British Heat, 29.
——— Heat, 29.
——— of Activity, 26.
——— of Activity, International, 26, 27.
Units of Work, 23.
Universal Welder, 211.

V.

Vegetable Food, Store-houses of Solar Energy, 14, 15.

INDEX. 289

Volt, or Unit of Electric Pressure, 32, 33.
—— Ampere or Watt, 36.
—— Coulomb-per-Second, 36.

W.

Wall Heater, Electric, 138.
Water, Circumstances Regulating Flow of, 37.
——, Conditions Requisite for Causing Flow of, 31.
—— Gramme-Degree-Centigrade, 29.
—— Heater, Electric, Low Economy of, 162, 163.
—— in Reservoir, Capacity of, for doing Work, 30.
—— Reservoir, Storage of Energy in, 30.
——, Resistance Offered by, to Escape of Heat from Conductors, 58, 59.
Watt, Definition of, 26, 27.
——, or Volt-Ampere, 36.
Welder, Automatic, 206–208.
——, Direct, 191–193.
—— for Cables, 213.
—— for Carriage Axle, 212.
—— for Carriage Tires, 210, 211.
—— for Shrapnel Shells, 217, 218.
—— for Wheel Spokes, 214.
——, Indirect, 197, 198.
——, Universal, 211.

Welding, Advantages Possessed by Alternating Currents in, 189.
—— Apparatus, Direct, 194–196.
——, Conditions Requisite for Obtaining Efficient Joints by, 183, 184.
——, Definition of, 182, 183.
——, Electric, 181–232.
——, Electric, Advantages Possessed by, 185, 186
——, Electric, Use of Continuous or Alternating Currents in, 187, 188.
—— Transformer, 201–206.
Wheel Spokes, Welder for, 214.
Wind, Storage of Solar Energy in, 16.
Wires, Bare Circuit, 64.
——, Covered Circuit, 64.
——, Fuse, 87–115
——, Safe Temperature Elevation of, 79.
Work, Definition of, 22.
—— done by Electric Current, 33.
——, Elements of, 22.
——, Units of, 23.
——, Unit of, Electric, 33.
Working of Electrical Metals, 259–270.

Z.

Zoroaster, 7.

Elementary Electro-Technical Series.

BY

EDWIN J. HOUSTON, Ph.D. and A. E. KENNELLY, D.Sc.

Alternating Electric Currents,
Electric Heating,
Electromagnetism,
Electricity in Electro-Therapeutics,
Electric Arc Lighting,
Electric Incandescent Lighting,
Electric Motors,
Electric Street Railways,
Electric Telephony,
Electric Telegraphy.

Cloth, profusely illustrated.   *Price, $1.00 per volume.*

The above volumes have been prepared to satisfy a demand which exists on the part of the general public for reliable information relating to the various branches of electro-technics. In them will be found concise and authoritative information concerning the several departments of electrical science treated, and the reputation of the authors, and their recognized ability as writers, are a sufficient guarantee as to the accuracy and reliability of the statements. The entire issue, although published in a series of ten volumes, is, nevertheless so prepared that each volume is complete in itself, and can be understood independently of the others. The books are well printed on paper of special quality, profusely illustrated, and handsomely bound in covers of a special design.

THE W. J. JOHNSTON COMPANY, Publishers,
253 BROADWAY, NEW YORK.

THIRD EDITION. GREATLY ENLARGED.

A DICTIONARY OF
Electrical Words, Terms, and Phrases.

By EDWIN J. HOUSTON, Ph.D. (Princeton).

AUTHOR OF

"Advanced Primers of Electricity"; "Electricity One Hundred Years Ago and To-day," etc., etc.

Cloth, 667 large octavo pages, 582 illustrations, Price, $5.00.

Some idea of the scope of this important work and of the immense amount of labor involved in it, may be formed when it is stated that it contains definitions of about 6000 distinct words, terms, or phrases. The dictionary is not a mere word-book; the words, terms, and phrases are invariably followed by a short, concise definition, giving the sense in which they are correctly employed, and a general statement of the principles of electrical science on which the definition is founded. Each of the great classes or divisions of electrical investigation or utilization comes under careful and exhaustive treatment; and while close attention is given to the more settled and hackneyed phraseology of the older branches of work, the newer words and the novel departments they belong to are not less thoroughly handled. Every source of information has been referred to, and while libraries have been ransacked, the notebook of the laboratory and the catalogue of the wareroom have not been forgotten or neglected. So far has the work been carried in respect to the policy of inclusion that the book has been brought down to date by means of an appendix, in which are placed the very newest words, as well as many whose rareness of use had consigned them to obscurity and oblivion.

Copies of this or any other electrical book published will be sent by mail, POSTAGE PREPAID, *to any address in the world, on receipt of price.*

The W. J. Johnston Company, Publishers,
253 BROADWAY, NEW YORK.

Electricity and Magnetism.
A Series of Advanced Primers.

By EDWIN J. HOUSTON, Ph.D. (Princeton).

AUTHOR OF

"*A Dictionary of Electrical Words, Terms and Phrases,*"
etc., etc., etc.

Cloth.   306 pages.   116 illustrations.   Price, $1.00.

During the Philadelphia Electrical Exhibition of 1884, Prof Houston issued a set of elementary electrical primers for the benefit of the visitors to the exhibition, which attained a wide popularity. During the last ten years, however, the advances in the applications of electricity have been so great and so widespread that the public would no longer be satisfied with instruction in regard to only the most obvious and simple points, and accordingly the author has prepared a set of new primers of a more advanced character as regards matter and extent. The treatment, nevertheless, remains such that they can be easily understood by any one without a previous knowledge of electricity. Electricians will find these primers of marked interest from their lucid explanations of principles, and the general public will find in them an easily read and agreeable introduction to a fascinating subject. The first volume, as will be seen from the contents, deals with the theory and general aspects of the subject. As no mathematics is used and the explanations are couched in the simplest terms, this volume is an ideal first book from which to obtain the preliminary ideas necessary for the proper understanding of more advanced works.

Copies of this or any other electrical book published, will be sent by mail, POSTAGE PREPAID, *to any address in the world, on receipt of price.*

The W. J. Johnston Company, Publishers,
253 BROADWAY, NEW YORK.

The Measurement of Electrical Currents and Other Advanced Primers of Electricity.

By EDWIN J. HOUSTON, Ph.D. (Princeton).

AUTHOR OF

"*A Dictionary of Electrical Words, Terms, and Phrases*," etc., etc., etc.

Cloth. 429 pages, 169 illustrations. Price, $1.00.

This volume is the second of Prof. Houston's admirable series of *Advanced Primers of Electricity*, and is devoted to the measurement and practical applications of the electric current. The different sources of electricity are taken up in turn, the apparatus described with reference to commercial forms, and the different systems of distribution explained. The sections on alternating currents will be found a useful introduction to a branch which is daily assuming larger proportions, and which is here treated without the use of mathematics. An excellent feature of this series of primers is the care of statement and logical treatment of the subjects. In this respect there is a marked contrast to most popular treatises, in which only the most simple and merely curious points are given, to the exclusion or subordination of more important ones. The abstracts from standard electrical authors at the end of each primer have in general reference and furnish an extension to some important point in the primer, and at the same time give the reader an introduction to electrical literature. The abstracts have been chosen with care from authoritative professional sources or from treatises of educational value in the various branches.

Copies of this or any other electrical book published will be sent by mail, POSTAGE PREPAID, *to any address in the world, on receipt of price.*

The W. J. Johnston Company, Publishers,
253 BROADWAY, NEW YORK.

THE
Electrical Transmission of Intelligence
AND OTHER ADVANCED PRIMERS OF ELECTRICITY.

By EDWIN J. HOUSTON, PH.D. (Princeton),

AUTHOR OF

"*A Dictionary of Electrical Words, Terms, and Phrases*," etc., etc., etc.

Cloth.   330 pages, 88 Illustrations.   Price, $1.00.

The third and concluding volume of Prof. Houston's series of *Advanced Primers of Electricity* is devoted to the telegraph, telephone, and miscellaneous applications of the electric current. In this volume the difficult subjects of multiple and cable telegraphy and electrolysis, as well as the telephone, storage battery, etc., are treated in a manner that enables the beginner to easily grasp the principles, and yet with no sacrifice in completeness of presentation. The electric apparatus for use in houses, such as electric bells, annunciators, thermostats, electric locks, gas-lighting systems, etc., are explained and illustrated. The primer on electro-therapeutics describes the medical coil and gives instructions for its use, as well as explaining the action of various currents on the human body. The interesting primers on cable telegraphy and on telephony will be appreciated by those who wish to obtain a clear idea of the theory of these attractive branches of electrical science and a knowledge of the details of the apparatus. Attention is called to the fact that each of the primers in this series is, as far as possible, complete in itself, and that there is no necessary connection between the several volumes.

Copies of this or any other electrical book published will be sent by mail, POSTAGE PREPAID, *to any address in the world, on receipt of price.*

The W. J. Johnston Company, Publishers,
253 BROADWAY, NEW YORK.

ELECTRICITY
ONE HUNDRED YEARS AGO AND TO-DAY.

By EDWIN J. HOUSTON, PH.D. (Princeton),

AUTHOR OF

"A Dictionary of Electrical Words, Terms, and Phrases," etc., etc., etc.

Cloth.   179 pages, illustrated.   Price, $1.00.

In tracing the history of electrical science from practically its birth to the present day, the author has, wherever possible, consulted original sources of information. As a result of these researches, several revisions as to the date of discovery of some important principles in electrical science are made necessary. While the compass of the book does not permit of any other than a general treatment of the subject, yet numerous references are given in footnotes, which also in many cases quote the words in which a discovery was first announced to the world, or give more specific information in regard to the subjects mentioned in the main portion of the book. This feature is one of interest and value, for often a clearer idea may be obtained from the words of a discoverer of a phenomenon or principle than is possible through other sources. The work is not a mere catalogue of subjects and dates, nor is it couched in technical language that only appeals to a few. On the contrary, one of its most admirable features is the agreeable style in which the work is written, its philosophical discussion as to the cause and effect of various discoveries, and its personal references to great names in electrical science. Much information as to electrical phenomena may also be obtained from the book, as the author is not satisfied to merely give the history of a discovery, but also adds a concise and clear explanation of it.

Copies of this or any other electrical book published will be sent by mail, POSTAGE PREPAID, to any address in the world, on receipt of price.

The W. J. Johnston Company, Publishers,
253 BROADWAY, NEW YORK.

PUBLICATIONS OF
THE W. J. JOHNSTON COMPANY.

The Electrical World. An Illustrated Weekly Review of Current Progress in Electricity and its Practical Applications. Annual subscription............. $3.00

The Electric Railway Gazette. An Illustrated Weekly Record of Electric Railway Practice and Development. Annual subscription 3.00

Johnston's Electrical and Street Railway Directory. Containing Lists of Central Electric Light Stations, Isolated Plants, Electric Mining Plants, Street Railway Companies—Electric, Horse and Cable—with detailed information regarding each ; also Lists of Electrical and Street Railway Manufacturers and Dealers, Electricians, etc. Published annually.... .... 5.00

The Telegraph in America. By Jas. D. Reid. 894 royal octavo pages, handsomely illustrated. Russia, 7.00

Dictionary of Electrical Words, Terms and Phrases. By Edwin J. Houston, Ph.D. Third edition. Greatly enlarged. 667 double column octavo pages, 582 illustrations...................... 5.00

The Electric Motor and Its Applications. By T. C. Martin and Jos. Wetzler. With an appendix on the Development of the Electric Motor since 1888, by Dr. Louis Bell. 315 pages, 353 illustrations........... 3.00

The Electric Railway in Theory and Practice. The First Systematic Treatise on the Electric Railway. By O. T. Crosby and Dr. Louis Bell. Second edition, revised and enlarged. 416 pages, 183 illustrations...... .... 2.50

Alternating Currents. An Analytical and Graphical Treatment for Students and Engineers. By Frederick Bedell, Ph.D., and Albert C. Crehore, Ph.D. Second edition. 325 pages, 112 illustrations................. 2.50

Publications of the W. J. JOHNSTON COMPANY.

Gerard's Electricity. With chapters by Dr. Louis Duncan, C. P. Steinmetz, A. E. Kennelly and Dr. Cary T. Hutchinson. Translated under the direction of Dr. Louis Duncan.................................... $2.50

The Theory and Calculation of Alternating-Current Phenomena. By Charles Proteus Steinmetz 2.50

Central Station Bookkeeping. With Suggested Forms. By H. A. Foster.......................... 2.50

Continuous Current Dynamos and Motors. An Elementary Treatise for Students. By Frank P. Cox, B. S. 271 pages, 83 illustrations................ 2.00

Electricity at the Paris Exposition of 1889. By Carl Hering. 250 pages, 62 illustrations. 2.00

Electric Lighting Specifications for the use of Engineers and Architects. Second edition, entirely rewritten. By E. A. Merrill. 175 pages.............. 1.50

The Quadruplex. By Wm. Maver, Jr., and Minor M. Davis. With Chapters on Dynamo-Electric Machines in Relation to the Quadruplex, Telegraph Repeaters, the Wheatstone Automatic Telegraph, etc. 126 pages, 63 illustrations... 1.50

The Elements of Static Electricity, with Full Descriptions of the Holtz and Topler Machines. By Philip Atkinson, Ph.D. Second edition. 228 pages, 64 illustrations..................................... 1.50

Lightning Flashes. A Volume of Short, Bright and Crisp Electrical Stories and Sketches. 160 pages, copiously illustrated................................. 1.50

A Practical Treatise on Lightning Protection. By H. W. Spang. 180 pages, 28 illustrations, 1.50

Publications of the W. J. JOHNSTON COMPANY.

Electricity and Magnetism. Being a Series of Advanced Primers. By Edwin J. Houston, Ph.D. 306 pages, 116 illustrations........................... $1.00

Electrical Measurements and Other Advanced Primers of Electricity. By Edwin J. Houston, Ph.D. 429 pages, 169 illustrations........ 1.00

The Electrical Transmission of Intelligence and Other Advanced Primers of Electricity. By Edwin J. Houston, Ph.D. 330 pages, 88 illustrations............................. 1.00

Electricity One Hundred Years Ago and To-day. By Edwin J. Houston, Ph.D. 179 pages, illustrated 1.00

Alternating Electric Currents. By E. J. Houston, Ph.D. and A. E. Kennelly, D.Sc. (Electro-Technical Series)................................. 1.00

Electric Heating. By E. J. Houston, Ph.D. and A. E. Kennelly, D.Sc. (Electro-Technical Series)...... 1.00

Electromagnetism. By E. J. Houston, Ph.D. and A. E. Kennelly, D.Sc. (Electro-Technical Series)...... 1.00

Electro-Therapeutics. By E. J. Houston, Ph.D. and A. E. Kennelly, D.Sc. (Electro-Technical Series).. 1.00

Electric Arc Lighting. By E. J. Houston, Ph.D. and A. E. Kennelly, D.Sc. (Electro-Technical Series).. 1.00

Electric Incandescent Lighting. By E. J. Houston, Ph.D. and A. E. Kennelly, D.Sc. (Electro-Technical Series)............................. .... 1.00

Electric Motors. By E. J. Houston, Ph.D. and A. E. Kennelly, D.Sc. (Electro-Technical Series)......... 1.00

Publications of the W. J. JOHNSTON COMPANY

Electric Street Railways. By E. J. Houston, Ph.D. and A. E. Kennelly, D.Sc. (Electro-Technical Series)..$1.00

Electric Telephony. By E. J. Houston, Ph.D. and A. E. Kennelly, D.Sc. (Electro-Technical Series).. 1.00

Electric Telegraphy. By E. J. Houston, Ph.D. and A. E. Kennelly, D.Sc. (Electro-Technical Series).. 1.00

Alternating Currents of Electricity. Their Generation, Measurement, Distribution and Application. Authorized American Edition. By Gisbert Kapp. 164 pages, 37 illustrations and two plates 1.00

Recent Progress in Electric Railways. Being a Summary of Current Advance in Electric Railway Construction, Operation, Systems, Machinery, Appliances, etc. Compiled by Carl Hering. 386 pages, 110 illustrations........................... 1.00

Original Papers on Dynamo Machinery and Allied Subjects. Authorized American Edition. By John Hopkinson, F.R.S. 249 pages, 90 illustrations.. 1.00

Davis' Standard Tables for Electric Wiremen. With Instructions for Wiremen and Linemen, Rules for Safe Wiring and Useful Formulæ and Data. Fourth edition. Revised by W. D. Weaver.......... 1.00

Universal Wiring Computer, for Determining the Sizes of Wires for Incandescent Electric Lamp Leads, and for Distribution in General Without Calculation, with Some Notes on Wiring and a Set of Auxiliary Tables. By Carl Hering. 44 pages... .... 1.00

Publications of the W. J. JOHNSTON COMPANY.

Experiments With Alternating Currents of High Potential and High Frequency. By Nikola Tesla. 146 pages, 30 illustrations.......... $1.00

Lectures on the Electro-Magnet. Authorized American Edition. By Prof. Silvanus P. Thompson. 287 pages, 75 illustrations........................ 1.00

Dynamo and Motor Building for Amateurs. With Working Drawings. By Lieutenant C. D. Parkhurst... 1.00

Reference Book of Tables and Formulæ for Electric Street Railway Engineers. By E. A. Merrill.................................... 1.00

Practical Information for Telephonists By T. D. Lockwood. 192 pages................... 1.00

Wheeler's Chart of Wire Gauges........... 1.00

A Practical Treatise on Lightning Conductors. By H. W. Spang. 48 pages, 10 illustrations. .75

Proceedings of the National Conference of Electricians. 300 pages, 23 illustrations......... .75

Wired Love ; A Romance of Dots and Dashes. 256 pages... .75

Tables of Equivalents of Units of Measurement. By Carl Hering............................ .50

Copies of any of the above books or of any other electrical book published, will be sent by mail, POSTAGE PREPAID, *to any address in the world on receipt of price.*

THE W. J. JOHNSTON COMPANY,
253 BROADWAY, NEW YORK.

THE PIONEER ELECTRICAL JOURNAL OF AMERICA.

Read Wherever the English Language is Spoken.

The Electrical World

is the largest, most handsomely illustrated, and most widely circulated electrical journal in the world.

It should be read not only by every ambitious electrician anxious to rise in his profession, but by every intelligent American.

It is noted for its ability, enterprise, independence and honesty. For thoroughness, candor and progressive spirit it stands in the foremost rank of special journalism.

Always abreast of the times, its treatment of everything relating to the practical and scientific development of electrical knowledge is comprehensive and authoritative. Among its many features is a weekly *Digest of Current Technical Electrical Literature*, which gives a complete *résumé* of current original contributions to electrical literature appearing in other journals the world over.

Subscription { including postage in the U. S., Canada, or Mexico, } **$3 a Year.**

May be ordered of any Newsdealer at 10 cents a week.

Cloth Binders for THE ELECTRICAL WORLD postpaid, $1.00.

The W. J. Johnston Company, Publishers,
253 BROADWAY, NEW YORK.

www.ingramcontent.com/pod-product-compliance
Lightning Source LLC
Chambersburg PA
CBHW031247250426
43672CB00029BA/1375